# Web GIS

## PRINCIPLES AND APPLICATIONS

PINDE FU
JIULIN SUN

ESRI PRESS
REDLANDS, CALIFORNIA

Esri Press, 380 New York Street, Redlands, California 92373-8100
Copyright 2011 Esri
All rights reserved. First edition 2011
15 14 13 12 11    2 3 4 5 6 7 8 9 10

Printed in the United States of America

*Library of Congress Cataloging-in-Publication Data*
Fu, Pinde, 1968–
    Web GIS : principles and applications / Pinde Fu, Jiulin Sun. — 1st ed.
      p. cm.
    Includes index.
    ISBN 978-1-58948-245-6 (pbk. : alk. paper) 1.  Geographic information systems. 2.  Geography
Computer network resources. 3.  World Wide Web. I. Sun, Jiulin. II. Title.
    G70.212.F8 2010
    910.285'4678—dc22                                              2010018972

Ask for Esri Press titles at your local bookstore or order by calling 1-800-447-9778. You can also shop online at www.esri.com/esripress. Outside the United States, contact your local Esri distributor.

Esri Press titles are distributed to the trade by the following:

*In North America:*

Ingram Publisher Services
Toll-free telephone: 1-800-648-3104
Toll-free fax: 1-800-838-1149
E-mail: customerservice@ingrampublisherservices.com

*In the United Kingdom, Europe, Middle East and Africa, Asia, and Australia:*

Eurospan Group
3 Henrietta Street
London WC2E 8LU
United Kingdom
Telephone: 44(0) 1767 604972
Fax: 44(0) 1767 601640
E-mail: eurospan@turpin-distribution.com

# TABLE OF CONTENTS

# PREFACE

The Web has revealed the immense value and applicability of GIS to e-government, e-business, e-science, industry, and daily life; set the bar for today's user expectations; and introduced flexible architectures for use with modern information technology infrastructure. Harnessing the power of Web GIS will greatly benefit, or even determine, the future of many organizations.

## OUR APPROACH

The demand for good Web GIS professionals is high. Colleges are responding to this need by establishing courses on Web GIS, and corporations depend on their staffs to take advantage of this technology. While the need is clear and pressing, there are not many Web GIS textbooks and learning guides available. This is due in large part to the challenges in this field—it is advancing too rapidly for books to keep up with, it is too new for many principles to gel, it is too diffuse to condense, and it is sometimes too technical to easily explain. Writing this book has been an adventure and a learning experience.

We have attempted to be all of the following:

- **Focused on the conceptual level:** Web GIS is a fast-moving field. Books that cover too many technical details tend to become quickly outdated and, in fact, require updates as soon as a new software version is released. We have tried to stay at the conceptual level, concentrating on methodology and principle, which is less transient and more likely to stay current for years to come. We do not include lab exercises for this reason, but instead provide suggestions for labs, which gives readers and instructors the freedom to use the latest resources online.

- **Comprehensive and systematic:** Web-GIS-related materials are scattered in various journals, wikis, blogs, and presentations. We have tried to organize these fragmented materials into a coherent and consistent entity, covering all the major forms of Web GIS, their architectures, development options, impacts on the GIS field, applications, issues or challenges, and prospects.

- **Novel:** We have tried to include the recent phenomena, research frontiers, and future trends of Web GIS, and we hope this book can broaden your horizons and help you to stay on top of Web GIS technology and applications.

- **Easy to understand:** We have been working on Web GIS applications and Web GIS research for more than a decade apiece. We have tried to convey our understanding of the field through plain language and deliberately designed graphics. We have included a range of examples drawn from real applications throughout the book to link Web GIS principles with practices and to help readers understand the real value of Web GIS to various industries.

## INTENDED AUDIENCE

We have attempted to design the book for readers at all skill levels and who have a variety of uses for GIS applications. Managers can learn the potential that Web GIS can bring to their organization; developers can find development options and best practices; government employees can learn how Web GIS can improve public services and enhance collaboration across agencies; businesses can learn how Web GIS can create new business models and reshape existing ones; and researchers can find new uses for Web GIS that will meet the challenges of the new frontier.

This book can be used as a college textbook. We have actually been using the contents while teaching a Web GIS course at the Environment and Planning College at Henan University, in Kaifeng, China, for more than two years. The course based on this book was well recognized by undergraduates, graduate students, and the university. We hope other instructors will find it to be an easy-to-adopt textbook for Web GIS courses in geography, environmental science, computer science, business, and other majors.

## SUGGESTED LABORATORY EXERCISES

We deliberately excluded lab exercises from this book. Including chunks of program source code and screen captures of step-by-step software tutorials can quickly outdate a book as soon as new software versions are released. However, we understand that many readers, especially college students, expect to learn Web GIS through hands-on exercises.

For our own teaching, we have designed the following series of lab exercises that progressively lead students to use Web GIS software and to build Web GIS solutions:

Lab 1    Find and assemble Web resources to build Web applications (using ArcGIS.com)
Lab 2    Develop a Web site (using HTML and JavaScript)
Lab 3    Understand REST Web services (with ArcGIS Server REST APIs)
Lab 4    Build mashup applications I (using ArcGIS API for JavaScript to integrate ArcGIS online services)
Lab 5    Create your own Web services using ArcGIS Server
Lab 6    Build mashup applications II (integrating your own services with other Web resources)
Lab 7    Optimize your Web map services (map caching)
Lab 8    Create and use 3D map services
Lab 9    Create and use geoprocessing Web services
Lab 10   Configure and extend ArcGIS Flex Viewer (optional)

Instructors and readers using ArcGIS Server can find abundant lab resources at `http://resources.esri.com/arcgisserver`. We have chosen ArcGIS Server because it is widely used and has the capability to serve the needs of many industries. However, instructors and readers can use other software options for teaching labs as well.

## SUMMARY

The lead author of this book works at ESRI, which has helped us to access a vast range of ESRI resources, from the latest Web GIS technology to a rich set of data, maps, and applications to the expertise of numerous colleagues. While in many cases we used ESRI products or applications as examples, the concepts and principles illustrated in this book apply to Web GIS in general and to all brands of products.

We welcome your feedback at `esripress@esri.com` and hope this book will spark your imagination and encourage creative uses of Web GIS.

Pinde Fu
Jiulin Sun
2010

# ACKNOWLEDGMENTS

This book wouldn't exist without many helping hands.

We would first like to thank Peter Adams, Judy Hawkins, and David Boyles of ESRI Press for their support and encouragement.

Our deep thanks also to Carolyn Schatz, ESRI Press editor. She greatly improved the language of the book, and her editing skills and familiarity with the intended audience helped us to find the best way to explain our ideas. Her attention to detail, dedication, patience, and hard work in coordinating the project were key to this book coming to fruition.

We'd like to express special appreciation to Professor Michael Goodchild of the University of California, Santa Barbara, and Derek Law of ESRI for carefully reviewing the book prior to publication; to Professors Paul Rich at the Creekside Center for Earth Observation LLC in Menlo Park, California, and Xiaojian Li at Henan University of Finance and Economics in Zhengzhou, China; and to Clint Brown, Ismael Chivite, Jeremy Bartley, Joseph Kerski, Marten Hogeweg, Roberto Lucchi, Jeff Shaner, Jennifer Jacob, and Anne Reuland of ESRI in Redlands, California, for reviewing individual chapters. Their outstanding suggestions and insights about Web GIS have guided us in enhancing the final product.

This book was developed based on our work and research experience at ESRI, the Institute of Geographic Sciences and Natural Resources Research (IGSNRR), the Chinese Academy of Sciences (CAS), and, very importantly, on the Web GIS course we taught at the Environment and Planning College at Henan University in Kaifeng, China. Deep thanks to Henan University, especially to Professors Yaochen Qin and Yunfeng Kong and Assistant Professors Yu Chen and Zhigang Han for their strong support and tremendous assistance. Deep thanks to our students for providing feedback that has helped us to improve the content and structure of this course.

We are grateful for the support of IGSNRR. Thanks to Jia Song, Chongliang Sun, Rui Li, Yuyue Xu, Xiuying Liao, Ling Yao, Runda Liu, Min Feng, Zhengchao Ren, Jinqu Zhang, and Yurui Xu at IGSNRR for their contributions to various chapters. The work of Jiulin Sun and IGSNRR colleagues was carried out with the financial support of the State Key Lab of Resource and Environment Information System of IGSNRR and the Data Sharing Network of Earth System Science within the National Scientific and Technology Infrastructure of China.

We are grateful for the support of ESRI. Thank you to Jack Dangermond for his strong support of this book and his inspired vision of Web GIS. Thanks to Bill Derrenbacher and Mourad Larif at ESRI for giving Pinde the flexibility to work on the book. We thank many of our colleagues for sharing enlightened discussions and for their support in many ways, particularly Ismael Chivite, Jeremy Bartley, Clint Brown, Urban MacGillivray, Christine Eggers, Jeff Shaner, Jian Lange, Gert van Maren, Moxie Zhang, Jay Chen, Bernard "Bernie" Szukalski, Jinwu Ma, Selim Dissem, Zichuan Ye, Dean Djokic, Denny Zhang, Michael Gould, Peter Eredics, Scott Morehouse, Victoria Kouyoumjian, Aileen Buckley, Satish Sankaran, Sterling Quinn, and ESRI China colleagues Xiaobing Cai, Zhenyu Li, Xin Chen, and Jingshun Guo.

It is amazing how many procedures and details are involved in producing a book. This book includes many case studies and screen captures, which include many content, data, and map providers. Acquiring permissions involved a considerable amount of work. We thank the team at ESRI Press for guiding us through the legal processes and for presenting our work in such a polished form. We also thank the organizations for their generous support in granting us the permission to use their materials.

While our names appear on the front cover of this book, our contributors' names deserve to be specially highlighted, too. We owe heartfelt thanks to our contributors Monica Pratt, Juanle Wang, Yunqiang Zhu, Fang Yin, and Yaochen Qin. Thanks for their diligent research of their chapter subjects and for working with us on numerous revisions.

Finally, we are grateful to our families. Pinde would like to thank his parents for buying him many books when he was young; his wife, Jun, for inspiring him to write; and his children, Kevin, Emily, and Vincent, for their patience while their father worked long hours. Jiulin would like to thank his wife, Xueying, for her consistent support and encouragement during the writing of this book.

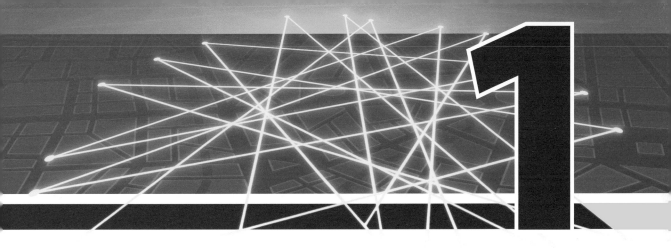

# GIS IN THE WEB ERA

The Web changes everything, and GIS is no exception. Web GIS, as the combination of the Web and geographic information system(s) or science (GIS), has grown into a rapidly developing discipline since its inception in 1993. GIS has turned into a compelling Internet application that has prompted many people to take advantage of the Web (Longley et al. 2005). The vast majority of Internet users use simple mapping or other spatially enabled applications over the Web at some point, though many are not aware of it. A few examples:

*Sitting comfortably at home, you can tour the world with the click of a mouse and appreciate the high-resolution satellite imagery that makes you feel like you're right there. Before you travel to a strange city, you can find the best hotels and examine the restaurants nearby. Web maps make a strange city as familiar as your hometown. When you are lost, you don't need to panic—your cell phone locates you on the map and GPS guides you back on track. Traffic congestion is not a worry—real-time traffic maps let you avoid traffic jams and get where you're going on time. When vacationing, you can blog about the places you go and map the photos you take to share with friends.*

*Government agencies map the transmission of infectious diseases, real-time earthquakes, and wildfire disasters online to keep members of the public healthy and safe. States and counties use online mapping to track where federal stimulus dollars are being spent to create a transparent government. Police officers gain situational awareness from real-time information reported by first responders in the field combined with information from other agencies so they can make sound decisions quickly in an emergency. Utility companies dispatch emergency maintenance orders to the nearest field employees and equip them with mobile maps of valves to shut down, spots to dig, pipes to fix, and driving directions for how to get there. Businesses, even small ones without a local GIS team, can still select the best store locations, understand their potential customers, and send them appropriate marketing messages. All of this is possible because of the benefits of GIS applications on the Web.*

GIS has benefited greatly from the Internet paradigm of broad connectivity and the momentum that the Web has generated. The Web has unlocked the power of GIS, from offices to laboratories. It has put GIS in the homes of millions and in the hands of billions, and made it usable across all industries, from government and business to education and research.

This chapter has four sections. Section 1.1 introduces the Internet, the World Wide Web, the mobile Web, and GIS. Section 1.2 presents early examples of Web GIS and how it has evolved in the Web 2.0 era. From these examples, section 1.3 summarizes the definition and characteristics of Web GIS. Section 1.4 then depicts the functions of Web GIS and its applications in e-business, e-government, e-science, and daily life. Lastly, this section uses the long tail theory to highlight the opportunities Web GIS offers in the mass market and in niche markets.

## 1.1 THE WEB AND GIS

The advent of the Internet and the Web are great milestones in the evolution of human civilization, as important as the invention of the printing press. The Internet and the Web helped pave the way for the information highway, allowing for an unprecedented information-based society and changing the way we live and work. Before introducing Web GIS, it is important to recognize the emergence and evolution of the Internet, the World Wide Web (WWW), the mobile Web, and GIS.

### 1.1.1 THE INTERNET, THE WEB, AND THE MOBILE WEB

During the Cold War in the 1960s, the U.S. Department of Defense Advanced Research Project Agency (ARPA) initiated a research project to create a network of geographically separated computers that could exchange information even if some of the nodes stopped functioning or were destroyed in the event of nuclear attack. In 1969, this research successfully connected the mainframes of four western universities—Stanford University; University of California, Santa Barbara; University of California, Los Angeles; and Utah State University (figure 1.1)—leading to the invention of the Internet. The network, ARPANet, was the predecessor of the global Internet (Internet Society 2003). New computers were gradually added to the network, including those of government agencies, universities, and research institutes, bringing the number of "nodes" to fifty-seven in 1975. The networks of these agencies are interconnected, thus forming the Internet. By the end of 1989, more than 100,000 computers were globally connected to the Internet.

The Internet didn't gain popularity until the 1990s. Before then, the services available over the Internet mainly consisted of e-mail, Usenet News (a way to provide news to Internet users before the Web was invented), file transfer, and Telnet (a protocol that allows you to log in to another computer and operate the computer remotely). The Internet was complex to use, its content was

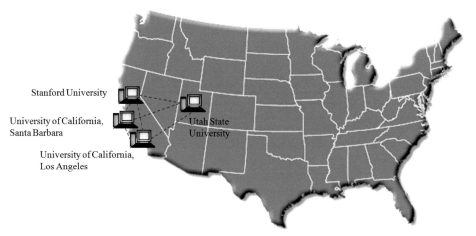

Stanford University

University of California,
Santa Barbara

Utah State
University

University of California,
Los Angeles

**Figure 1.1** ARPANet, which successfully connected the mainframes of four western universities, is recognized as the predecessor of the global Internet.

not nearly as rich as it is today, and its users were mostly professionals from research institutes and government agencies.

In 1990, Tim Berners-Lee, a researcher at CERN (European Organization for Nuclear Research), dramatically changed the way the Internet was used. While trying to find an easy way to share and exchange documents with his colleagues, Berners-Lee invented HTTP (Hypertext Transfer Protocol), HTML (Hypertext Markup Language), and the URL (Uniform Resource Locator). He developed the world's first Web server and a Web browser, naming his invention the World Wide Web. With the birth of the World Wide Web, Berners-Lee is undisputedly recognized as the "father of the Web" (figure 1.2).

The Web has made the Internet interesting, easy, and fun to use; it has forever changed the way we live and work; and it has changed the computer's role from calculation to one of communication and entertainment. The type of contents, as indicated by the number of Web sites available over the Internet, has grown exponentially, as the number of Internet users and the Web's penetration have

**Figure 1.2** Tim Berners-Lee, who invented the URL, HTTP, and HTML, is recognized as the "father of the Web."
Courtesy of CERN.

risen steadily (International Telecommunication Union 2010; Netcraft 2010). You can surf the Web rather than read the newspaper. You can send dozens of e-mails everyday instead of using the post office. You can search for the best deal via online shopping and auction Web sites such as eBay and Amazon rather than hitting all the malls. You can catch up with old friends and make new ones using social networking sites such as Facebook and Twitter instead of traveling back and forth. You can watch video on YouTube instead of on TV. You can chat with your cyber friends using instant messaging without worrying about the phone bill. You can share thousands of photos with relatives using free online photo albums instead of paying to develop and mail them. You can discover local businesses with a mouse click without bothering to flip through the phone book. You can conduct video conferences comfortably from your office rather than having to wake up early and catch the red-eye. All of this has become so natural that many of us, especially those who grew up with the Web, can't imagine life without it.

The terms Internet and the World Wide Web are synonymous in the minds of many, but they have different meanings. **The Internet is a massive network of networks that connects millions of computers worldwide.** Computers connected to the Internet can communicate with one another with a number of protocols such as HTTP, SMTP (Simple Mail Transfer Protocol), FTP (File Transfer Protocol), IRC (Internet relay chat), IM (instant messaging), Telnet, and P2P (peer-to-peer). **The World Wide Web is a system of interlinked hypertext documents and programs that can be accessed via the Internet primarily by using HTTP.** While HTTP is only one of the protocols the Internet supports, the Internet's chief attraction for a large number of users is the content accessible on the Web and the activities the Web facilitates. So it is that **the Web is the "face" of the Internet** (Douglas 2008).

The Web is still quickly expanding, from wired networks into wireless networks, fed by the popularity of the mobile Web, or wireless Web. The advances of mobile devices such as smart phones and wireless communications technologies, including wi-fi (wireless fidelity) and the 3G (third-generation) cellular network, have facilitated its spread. The International Telecommunication Union (2010) expected the number of mobile cellular subscriptions in 2010 to reach 5 billion globally, which is more than half the world's population. More and more of these subscribers are using their cell phones to connect to the Internet and surf the Web. **The wireless Internet will undoubtedly grow several times bigger than the wired Internet, and the mobile Web will give people the freedom to surf the Web on the go, anywhere, anytime.**

### 1.1.2 GIS

Everything that happens, happens somewhere. Knowing "what" is "where," and "why" it is there, can be critically important for making decisions in personal life as well as in an organization. GIS is the technology as well as the science for handling the "where" type of questions and for making intelligent decisions based on space and location.

**GIS is a system of hardware, software, and procedures that capture, store, edit, manipulate, manage, analyze, share, and display georeferenced data.**

GIS technology has been around since before the Internet and the Web. The first operational GIS was developed in 1962 by Roger Tomlinson (figure 1.3) for Canada's Federal Department of Forestry and Rural Development. Called the Canada Geographic Information System (CGIS), it was

**Figure 1.3**   Roger Tomlinson, who developed the first computerized GIS in the world in 1962, coined the term "geographic information system" and is recognized as the "father of GIS."

used for Canadian land inventory and planning. Tomlinson has become known as the "father of GIS" for his pioneering work developing CGIS and promoting GIS methodology (Tomlinson 2008).

GIS can produce more than just pretty maps, although GIS is conventionally used to make a myriad of maps using different scales, themes, and symbols. More importantly, GIS has powerful analytical functions that turn data into useful information. GIS can relate otherwise disparate data on the basis of common geography, revealing hidden relationships, patterns, and trends that are not readily apparent in spreadsheets or statistical packages, and create new information that can support informed decision making. For instance, as illustrated in figure 1.4, the real world can be abstracted into a number of spatial data layers, including land use, elevation, imagery, parcels, streets, and business customers. These layers can certainly be used to create composite maps, but they can also be used to generate a variety of useful information through GIS analysis.

This information can be used to answer such crucial questions as the following:

- If there is a flood, what areas will be at risk? And whose homes will be affected?

    - Analysis: This can be roughly estimated with a buffer analysis (e.g., 50 meters from the river side) or, more accurately, with a 3D surface volume analysis that accounts for elevation or a more advanced flood simulation model that accounts for the dynamic nature of rainfall and water flow.

    - Use: A city can use this information for zoning and planning (e.g., the residential area should avoid areas exceeding a certain level of flood risk) and emergency preparedness (which areas should be evacuated in case of a flood).

- Which customers might be affected?

    - Analysis: A business can use GIS to overlay the aforementioned risk area with the customer layer to determine which customers' houses fall into the flood zone.

    - Use: An insurance company can calculate the total financial impact from a potential catastrophic flood.

- What is the action plan in case of a flood?

    - Analysis: GIS can be used to find the areas upstream where the river bank can be altered to redirect the floodwaters to an unpopulated area.

    - Use: Emergency crews can make preparations to reduce damages and save lives.

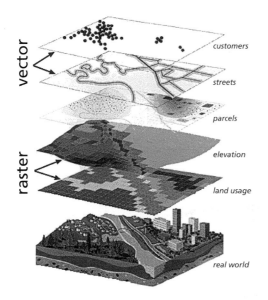

**vector**

customers

streets

parcels

**raster**

elevation

land usage

real world

**Figure 1.4** GIS abstracts the real world into data models consisting of multiple layers, where each layer represents a different theme of real-world data. This data can be presented in 2D or 3D maps, and GIS can be used to perform analysis to solve real-world problems such as finding the areas that are at high flood risk, what the total property damage would be, and what alternative plans can be made to mitigate the risk.

The preceding example is overly simplified. A rich set of data management, visualization, and analysis capabilities has been developed by GIS professionals since 1962, making GIS an essential tool for land-use planning, utilities management, ecosystems modeling, crime analysis, market analysis, tax assessment, and many other applications. **GIS capabilities go beyond mapping. GIS offers a rich set of analytical functions that can reveal hidden relationships, patterns, and trends that are not readily apparent, enabling people to think spatially to solve problems and make smart decisions.** GIS is the supporting science and technology for GeoDesign, which is a systematic methodology for geographic planning and decision making. The GeoDesign application of GIS (figure 1.5) can help people understand and analyze the world's problems and design alternatives that can lead the world to a better future (Steinitz 1990; Dangermond 2009a).

For decades, GIS professionals have used GIS technology to integrate, analyze, and visualize geographic information and knowledge, leading to abundant GIS applications benefiting many fields. However, GIS still has great potential that has not been fully realized. Access to GIS has been limited to a relatively small number of GIS professionals. The emergence of Web GIS is unlocking the power of GIS to a wider audience. The Web has made GIS not only more accessible to people in their offices, homes, and on the go, but also more flexible through Web-based APIs. A developer uses an API (application programming interface) to facilitate seamless integration with other information systems.

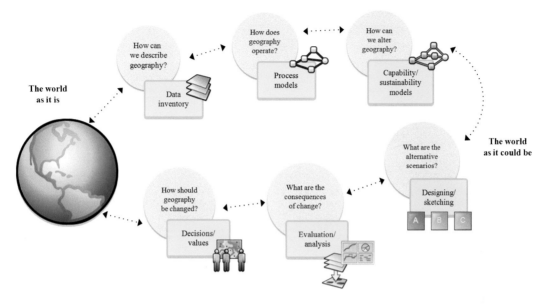

**Figure 1.5** GIS is the foundation science and technology for GeoDesign, a systematic methodology of geographic planning and decision making.

## 1.2 WEB GIS ORIGINS AND EVOLUTION

The fusion of the Internet and the Web and traditional disciplines has created many new ones, and Web GIS is one of these disciplines. Web GIS has evolved rapidly since 1993, especially in the so-called "Web 2.0" era. Web GIS has considerably changed the way geospatial information is acquired, transmitted, published, shared, and visualized. It represents a significant milestone in the history of GIS.

### 1.2.1 EARLY WEB GIS

In 1993, the Xerox Corporation Palo Alto Research Center (PARC) developed a Web-based map viewer, marking the origin of Web GIS. The Xerox PARC Map Viewer was an experiment in allowing retrieval of interactive information on the Web, rather than providing access to strictly static files (Putz 1994). The Web site provided simple map zoom capabilities, layer selection, and map projection conversion functions. Users could use the map viewer in a Web browser and click a link to a function. The Web browser would then send an HTTP request to the Web server. The Web server would receive the request, perform mapping operations, generate a new map, and return it to the Web browser that requested it. The Web browser would then receive and display the map image. This pioneered the approach of running GIS inside a Web browser, demonstrating that users anywhere on the Web could use GIS without having it locally installed, an advantage that traditional desktop GIS does not have.

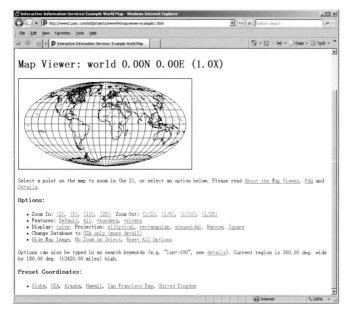

**Figure 1.6** The PARC Map Viewer is recognized as the first Web GIS application. It allowed people to use basic map functions from anywhere with an Internet connection without having to install GIS software and data locally.

Courtesy of Steven B. Putz.

Realizing its benefits, the GIS community quickly adopted this concept of using GIS functions in Web browsers. Subsequently, numerous Web GIS applications have emerged. Here are a few examples:

- In 1994, the Canadian National Atlas Information Service released the first online version of the National Atlas of Canada. It is an interactive mapping Web site that allows the public to select a number of data layers such as roads, rivers, administrative boundaries, and ecological regions, and then submit the map request. The server can choose the appropriate symbol to generate maps on request, and citizens can view the map atlas online at home without having to go to a government office to see the results.

- In 1995, the University of California, Santa Barbara, leading a number of other organizations, developed the Alexandria Digital Library (Frew et al. 1995); and the U.S. Geological Survey (USGS) implemented a Web site portal for the National Geospatial Data Clearinghouse (Nebert 1995). These two Web applications allow Web users to specify keywords and an area on the map, and then search for maps and satellite images that match these criteria. These applications facilitate sharing geospatial information, an early example of geoportals (see chapter 6).

- In 1995, the U.S. Census Bureau released its TIGER (Topologically Integrated Geographic Encoding and Referencing) Mapping Service that allows the public to query and map the demographic information of states, counties, and cities. The TIGER Mapping Service delivers the vast volume of information that is available in the national census database in online map format to citizens' Web browsers.

- Susan Huse (1995) at the University of California, Berkeley, developed GRASSLinks as part of her PhD dissertation. GRASS (Geographic Resources Analysis Support System) was a desktop GIS tool then, and its functions were not exposed to the Web. Huse implemented an interface between the Web server and GRASS that allowed users to select data layers from their Web browser and submit requests to the Web server. The server forwarded the request to GRASS, where buffer, overlay, reclassification, and mapping operations were taking place, and the results were returned to users. GRASSLinks was an early example that demonstrated that Web GIS could go beyond mapping and query to perform sophisticated analysis.

- In 1996, MapQuest released its Web mapping application. It allowed people to view maps, look for local businesses, find the optimum route to a desired location, and plan trips. It represents an early precedent for the consumer mapping Web sites that are popular today.

PARC Map Viewer (figure 1.6) and many of the early Web GIS applications provided only limited functionality, and some couldn't perform well when there were many concurrent users. Nevertheless, they clearly demonstrated the benefits of using the Web as a medium to deliver GIS functions to a wide audience. Those who use it don't have to sit by the computer where the GIS application is installed. They can be anywhere, even on the other side of the earth, as long as there is an Internet connection. They can visit a Web site using a freely available Web browser without having to pay anything. They can use the basic GIS mapping functions provided by the map viewer without needing to install GIS software and data on a local machine. GIS companies began releasing their commercial Web GIS software in 1996. These commercial products are used by a variety of organizations, including government agencies such as USGS, the U.S. Environmental Protection Agency (EPA), Department of Housing and Urban Development (HUD), and the Bureau of Land Management (BLM), to build applications in many fields.

## 1.2.2 WEB GIS IN WEB 2.0

The term "Web 2.0" was first coined by Darcy DiNucci (1999), although it is now closely associated with Tim O'Reilly because of the O'Reilly Media Web 2.0 conference in 2004 and because of an article he wrote expounding on the subject (2005). Although the term suggests a new version of the World Wide Web, it does not signify an update of any technical specifications, but rather refers to cumulative changes in the ways software developers and end users use the Web.

In his "What is Web 2.0" article, O'Reilly pointed out that the bursting of the dot-com bubble in the fall of 2001 marked a turning point for the Web—far from having "crashed," the Web was more important than ever, with exciting new applications and Web sites popping up with surprising regularity. Among these new features are the following:

- **Harnessing collective intelligence:** Web 1.0, the original Web, was characterized by read-only content and a top-down information flow. Web 2.0 is a read-write Web that features an abundance of user-generated content (UGC) and a reverse (i.e., bottom-up) information flow. The competitive advantage derived by some Web 2.0 companies comes almost entirely from the critical mass of users sharing information—for example, eBay's product is the collective activity of all its users buying and selling goods and services; Amazon engages users by inviting them to submit and share product reviews; blog Web sites are based on users' posts and diaries; Wikipedia, an online encyclopedia, is made up of the expertise of Web users and is edited by volunteers. These Web sites attract new and regular users because of the products, comments, and knowledge offered by other users.

- **The Web as a platform:** The Web is a platform for computing and software development. Among these developments are Software as a Service (SaaS), where software capabilities are delivered as Web services (see chapter 3) or Web applications, and cloud computing, where dynamically scalable and often virtualized resources are provided as a service over the Web (see chapters 7 and 10). These Web services can be remixed "in the cloud" to build new applications. The Web as a platform makes these new methods available for application development and deployment.

- **Lightweight programming models:** Many Web systems support lightweight and easy programming models such as AJAX (Asynchronous JavaScript and XML) and empower users to create new functionality with mashups simply by mashing up, or assembling, various Web sources in novel ways.

- **Data is the next "inside intel":** Every significant Internet application is backed by a specialized database. Data is important, especially when there is significant cost in creating the data. Companies can license unique data or aggregate data, for example, by facilitating user participation. Once the amount of data reaches critical mass, it can be turned into a system service.

- **Software that reaches beyond a single device:** Web applications can be accessed by an increasing variety of devices, including Web browsers, desktop clients, and diversified cell phones, stemming from advances in wireless communications.

- **Rich usability:** Popular Web 2.0 applications have a graphic user interface that is easy to use, is easy on the eye, and offers a user experience that is similar to what desktop software applications have to offer (see chapters 2 and 4).

Equation 1.1 is a short summary of the main characteristics of Web 2.0 that make it more interactive, more integrated, and more useful to its users. The geospatial industry, including both consumer Web mapping companies and professional GIS companies, seeks to follow the principles of providing a rich user experience, encouraging user participation, and offering lightweight APIs so users can create their own applications (Maguire 2008).

---

**EQUATION 1.1**

```
Web 2.0 = User-generated content + the Web as a platform +
          a rich user experience
```

---

Commercial Web mapping applications such as Google Maps, Google Earth, Microsoft Bing Maps (formerly Virtual Earth), Yahoo Maps, and MapQuest are usually considered good examples of Web 2.0. These sites commonly provide detailed maps and high-resolution ground imagery of many regions. Street View by Google and StreetSide by Microsoft present 360° horizontal and 290° vertical panoramic photos along the streets of many regions. The Bird's Eye aerial photos by Microsoft and Oblique Aerial Imagery by Google also offer great ground detail. Signs,

**Figure 1.7** Consumer Web maps such as Microsoft Bing Maps for Mobile provide location-based services for people on the go.

Data courtesy of Microsoft, NAVTEQ, Harris Corp., Earthstar Geographics LLC, EarthData, Getmapping plc, AND Automotive Navigation Data, GeoEye, MapData Sciences Pty Ltd, NASA, and U.S. Geological Survey.

advertisements, pedestrians, and other objects are clearly visible. These detailed images have stimulated a surge of interest in the world around us, letting you visit famous landmarks and scenic sites; fly across mountains and lakes; view urban skyscrapers; take a virtual walk through places you want to go; and then search for the best route to take, restaurants to try, and places to stay. The user interfaces are intuitive, dynamic, responsive, and rich—like a good video game in many ways. These sites provide maps for mobile devices and provide services based on your location, allowing you to find the restaurants and hotels nearby (figure 1.7).

Professional GIS companies have adopted Web 2.0 principles and design patterns in Web GIS product lines to facilitate the sharing, communication, interoperability, collaboration, and integration of geospatial information on the Web. The ESRI product line, for example, embodies these principles, which include the following:

- **Harnessing of collective intelligence and data as the next "inside intel":** The geospatial Web services, especially feature editing services, and mashup capabilities provided by ArcGIS Server allow organizations to collect and share geographic knowledge, promoting collaboration among the geospatial community. The ArcGIS.com and ArcGIS community maps program provide a platform for organizations to share their data, maps, and applications. The data is aggregated, sometimes augmented, published as services, and then hosted by ESRI over the Web. This content can be accessed by contributing organizations and designated user groups, as well as by the public at large.

- **Using the Web as a platform:** Web services are the basic programming components of the Web platform. ESRI ArcGIS Server allows organizations to publish their authoritative basemaps, globes, and geoprocessing functions as Web services (see chapter 3). ArcGIS.com and ArcGIS Online provide cloud-based software and services, hosted user storage, and access to GIS tools and imagery to the GIS community (see chapters 7 and 10). These geospatial services have standard interfaces that can be easily remixed. This opens up opportunities for organizations to use the Saas and S + S (Software plus Service) approaches to reduce costs and increase system flexibility.

- **Mashup-style programming:** Web services from multiple agencies can be easily integrated, or mashed up, using the ArcGIS REST API (see chapter 3) or lightweight ArcGIS APIs for JavaScript, Adobe Flex, and Microsoft Silverlight (see chapter 4). Building powerful Web GIS applications to accomplish enterprise workflows becomes simple, quick, and dynamic.

- **Mobile solutions:** ESRI puts GIS in a range of mobile platforms, including Apple iPhone, Google Android, and Research in Motion BlackBerry. Users can retrieve data, view maps, and use analytical models directly from ArcGIS Server. The mobile client can also post geospatial data collected or verified in the field along with field photos and videos directly to ArcGIS Server. Server data is updated in real time to support sound and quick decision making in the office (see chapter 5).

- **A rich user experience:** Developers can create rich Web GIS applications with the use of ArcGIS APIs for JavaScript, Flex, and Silverlight. The interface integrates the effects of multimedia and animation to improve user satisfaction and increase productivity (see chapters 2 and 4). ArcGIS Explorer, as a 2D map viewer and a 3D virtual globe, is easy and fun to use. It can display terrain and other themes in 3D, allowing users to explore and discover geospatial patterns that would not be easily visible in 2D. With a few mouse clicks, ArcGIS Explorer can also combine diverse Web services and dynamic feeds with users' local data to create a common operational picture. It can also perform advanced analysis by calling ArcGIS Server behind the scenes (figure 1.8).

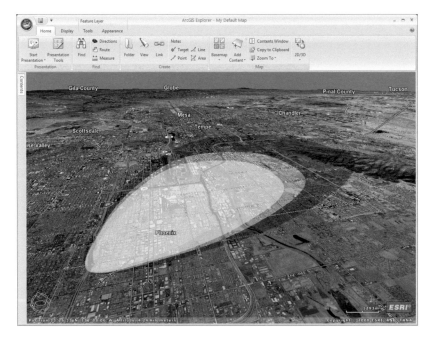

**Figure 1.8**  ArcGIS Explorer Desktop offers both a 2D map viewer and a 3D virtual globe. It also provides an interface to ArcGIS Server geoprocessing models. Here, ArcExplorer calls a plume model exposed in an ArcGIS Server geoprocessing service to show the areas that will be affected by a release of toxic gas.

Map data © AND Automotive Navigation Data; courtesy of Tele Atlas North America, Inc.

The rich features offered by GIS companies are relatively new, but users are rapidly learning the skills needed to take advantage of these new services. Many examples will be provided throughout the book to show how these applications can be used in a variety of fields, including e-government, e-business, academic research, and personal life.

## 1.3 WEB GIS CONCEPT

Web GIS started off as GIS running in Web browsers and has evolved into Web GIS serving desktop and mobile clients in addition to Web browser clients. This section will examine the definition of Web GIS, underscoring its characteristics as compared with traditional desktop GIS.

### 1.3.1 DEFINITION

Web GIS is a type of distributed information system. The simplest form of Web GIS should have at least a server and a client, where the server is a Web application server, and the client is a Web browser, a desktop application, or a mobile application (figure 1.9). The server has a URL so that clients can find it on the Web. The client then relies on HTTP specifications to send requests to the server. The server performs the requested GIS operation and sends a response to the client, again via HTTP. The format of the response can be an HTML that is used by the Web browser client, but it can also be in other formats such as binary image, XML (Extensible Markup Language), or JSON (JavaScript Object Notation) (see chapter 2).

Web GIS is often thought of as GIS running in a Web browser, but this definition overlooks systems with desktop clients and mobile clients. **Web GIS is any GIS that uses Web technologies. In a narrower definition, Web GIS is any GIS that uses Web technology to communicate between components.**

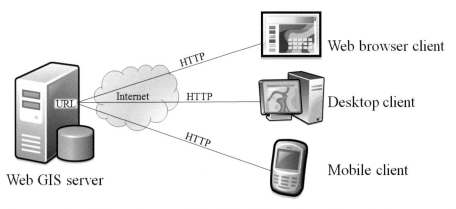

**Figure 1.9** The simplest architecture of Web GIS should have at least a Web application server and a client, which can be a Web browser, desktop client, or mobile client. The server and the client communicate via HTTP.

Web GIS is further defined by the following:

- HTTP, among many Web technologies, is the main protocol used by the different components of Web GIS to communicate with each other. If the Internet is down, some thick clients (see chapters 2 and 5) can still function based on caches and preloaded data or functions, but most clients will stop functioning.

- The simplest architecture of Web GIS is a two-tier system that involves a server and one or more clients. A Web GIS is not just the programs running on your computer, but more importantly, the server residing somewhere on the Web, or "in the cloud." Sometimes, the server and the client can run on one computer, but they are actually two separate components.

- Many Web GIS architectures consist of three tiers, including a data tier. And now as the mashup approach extends the reach of Web services, Web GIS is increasingly becoming more than three tiers (see chapter 4). These tiers and components can be distributed to a variety of locations over the Internet.

- Web GIS and desktop GIS are increasingly intertwined. Web GIS relies on desktop GIS to author resources. Desktop GIS, on the other hand, has expanded its functionality to make use of the resources on the Web. For instance, as an ArcGIS desktop user, you can use basemaps available over the Web, such as those served by USGS or Microsoft Bing Maps, without having to have your own copy of the data on your local computer.

Web GIS is closely related to two other terms: Internet GIS and the geospatial Web. Internet GIS (Peng and Tsou 2003) and Web GIS are often used synonymously. Strictly speaking, however, the two are slightly different. The Internet supports many services, and the Web is only one of them. GIS that uses any of the Internet services, and not just the Web, can be considered Internet GIS—making Internet GIS theoretically broader than Web GIS (figure 1.10). In reality, the Web is the chief attraction of the Internet and is the most commonly used Internet service. Thus, Web GIS is the most pervasive form of Internet GIS.

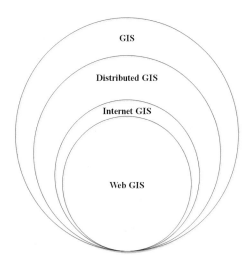

**Figure 1.10** Web GIS is shown in relation to other related GIS terms. While Internet GIS has a slightly broader coverage, Web GIS is the most widely used form of Internet GIS. Thus, many people use the terms synonymously.

The geospatial Web, or GeoWeb, is another term used to refer to Web GIS. However, the definition of GeoWeb is not identical to Web GIS. One definition of the GeoWeb is the merging of geospatial information with abstract (nongeospatial) information (e.g., Web pages, photos, videos, and news) that currently dominates the Web (Haklay, Singleton, and Parker 2008). This definition is closely related to the geotagging and geoparsing research areas of Web GIS (see chapter 10). Others use GeoWeb to refer to the emerging distributed global GIS, which is a widespread distributed collaboration of knowledge and discovery that promotes and sustains worldwide sharing and interoperability (Scharl and Tochtermann 2007; Dangermond 2009b). This definition highlights the contributions that Web GIS can make and illustrates the potential for Web GIS to grow.

A GIS basically consists of hardware, software, data, and users. It is interesting to see how the distance between these physical components has increased over the years. In the 1960s and 1970s, these components were usually colocated on one computer. In the 1980s and early 1990s, distributed GIS emerged with the adoption of a local area network (LAN). These components no longer needed to be colocated and were commonly located separately in one or multiple buildings of an organization. With the birth of Web GIS, these components are now separated even further apart. Most notably, GIS users sitting on one side of the globe can access a server located on the other side of the globe. With the emergence of newer technologies such as Web services (see chapter 3) and mashups (see chapter 4), a client can use data from one server and functions from another. The server of one client can be the client of another server, further separating and distributing the components of a GIS.

## 1.3.2 CHARACTERISTICS

The Internet and the Web removed the constraint of distance from cyberspace, allowing instant access to information over the Web without regard to how far apart the user and the server might be from each other. This quality gives Web GIS inherent advantages over traditional desktop GIS, including the following:

- **A global reach:** A developer can present Web GIS applications to the world, and the world can see them. A user can access Web GIS applications from their home computer or cell phone. The global nature of Web GIS is inherited from HTTP, which is broadly supported. Almost all organizations open their firewalls at certain network ports to allow HTTP requests and responses to go through their local network, increasing accessibility.

- **A large number of users:** In general, a traditional desktop GIS is used by only one user at a time, while a Web GIS can be used by dozens or hundreds of users simultaneously. This requires much higher performance and scalability for Web GIS than for desktop GIS.

- **Better cross-platform capability:** The majority of clients of Web GIS are Web browsers. There are Web browsers such as Internet Explorer, Mozilla Firefox, Apple Safari, and Google Chrome for diverse operating systems, including Microsoft Windows, Linux, and Apple Mac OS. Because these Web browsers largely comply with HTML and JavaScript standards, Web GIS that relies on HTML clients typically supports different operating systems. Web GIS that relies on Java, .NET, and Flex clients can run on multiple platforms where the required run-time environment is installed. It is worth mentioning that Web GIS for mobile clients is far from being cross platform because of the diversity in mobile operating systems and the current incompatibility of mobile Web browsers.

- **Low cost as averaged by the number of users:** In the spirit of the Internet, the vast majority of Internet contents are free of charge to end users, and this is true of Web GIS. Generally, you do not need to buy software or pay to use Web GIS. Organizations that need to provide GIS capabilities to many users can also keep their Web GIS costs low. Instead of buying and setting up desktop GIS for every user, an organization can set up just one Web GIS, and this single system can be shared by many users—from home, at work, or in the field. The reduced costs in purchasing and maintenance help to provide a high return on investment.

- **Easy to use for end users:** Desktop GIS is intended for professional users with years of training and experience in GIS. Web GIS is intended for a broad audience, including public users who may know nothing about GIS. They expect Web GIS to be as easy as using a regular Web page. In the Web 2.0 era, their expectations are even higher—"if I don't know how to use your site, it's your fault." So, Web GIS is commonly designed for simplicity, intuition, and convenience, making it typically much easier for end users than desktop GIS.

- **Unified update:** For desktop GIS to be updated to a new version, the update needs to be installed on every computer. For Web GIS, one update works for all clients, making updating a lot easier. As long as the program and data are updated on the server, most Web GIS clients will get automatic updates. This means easy maintenance for Web GIS and greatly improved timeliness for GIS, making Web GIS a good fit for delivering real-time information.

- **Diverse applications:** Unlike desktop GIS, which is limited to a certain number of GIS professionals, Web GIS can be used by everyone in an enterprise as well as by the public at large. This broad audience has diverse demands, which results in Web GIS being used in a variety of applications, both formal and informal. For example, a Web site in Australia delivers the locations of public toilets (`http://www.toiletmap.gov.au`). **Neogeography (i.e., "new geography") is a concept that is gaining popularity as nonexpert users employ geographic techniques and tools for personal and community purposes** (Turner 2006). While not strictly Web-based, neogeography typically refers to such simple and informal applications as mapping celebrity homes, tagging personal photos, locating friends, displaying Wi-Fi hot spots, and marking locations of news events.

These characteristics reveal both the advantages and challenges facing Web GIS. For example, the easy-to-use nature of Web GIS stimulates public participation, but it also reminds Web GIS designers to take into account Internet users and officials who have no GIS background. The large number of users supports greater GIS deployment, but it also requires that Web GIS be scalable—that is, able to maintain good performance as the number of users increases.

## 1.4 WEB GIS APPLICATIONS

Web GIS can deliver all GIS functions over the Web, and these functions can, in turn, be applied in a wide range of industries. Still, furthering the applications of Web GIS is far from reaching its full potential, which represents a great opportunity for GIS professionals to seize.

### 1.4.1 FUNCTIONS

Web GIS is able to perform all GIS functions involving spatial information, including capture, storage, editing, manipulation, management, analysis, sharing, and visualization. Some of the strengths of Web GIS include the following:

- **Mapping (visualization) and query:** Web mapping, as the face of Web GIS, is the most commonly used function. GIS data and analysis results are usually presented as maps. Every location on the map, in turn, has attributes that support such operations as spatial identify (i.e., what is located here?) and attribute query (i.e., where are the bookstores located?).

- **Collection of geospatial information:** There has been a great interest, by amateurs as well as professionals, in using the Web to create and assemble a trove of geographic information. For example, as of January 2010, the Wikimapia Web site has nearly 12 million places voluntarily marked and described across the globe by registered and anonymous Internet users. OpenStreetMap is a collaborative project based on users who collect data using personal portable GPS devices. Michael F. Goodchild (2007) termed such information VGI (volunteered geographic information, see chapter 10).

While VGI is useful, many types of decisions need to be made from verified, authoritative content to ensure reliability and accuracy. Industry enterprises and professionals have adopted Web GIS to collect authoritative information from a variety of sources. With the use of a mobile GIS client, field data can be collected and validated by field crews and posted to a server and database in the office for timely information updates (see chapter 5). Web feature editing services enable sophisticated data editing in Web browsers.

- **Dissemination of geospatial information:** Web GIS is an ideal platform for the wide distribution of information. The VGI Web sites mentioned earlier allow users to view operational results and download data. Government agencies, the academic sector, and some commercial sectors have long used Web GIS to share geospatial information (see chapter 6). For example, the National Geospatial Data Clearinghouse and Alexandria Digital Library project in the 1990s built Web-based platforms to facilitate geospatial data sharing, discovery, evaluation, and download. More recent GIS Web portals such as the U.S. Geospatial One-Stop (GOS) portal, European Union Infrastructure for Spatial Information in Europe (INSPIRE), and ArcGIS.com not only allow users to search and download data, but also to use the geospatial Web services live. These geoportals encourage collaboration and cooperation among and across departments and organizations. They help organizations to leverage existing geospatial resources rather than duplicate efforts re-creating them, leading to reduced costs and increased efficiency.

- **Geospatial analysis:** Web GIS has gone beyond mere mapping. It also provides analytical functions—most notably, those closely related to daily life, such as measuring distances and areas, finding the optimum driving path (i.e., navigation), finding the location of an address or place, and using proximity analysis to find the businesses nearby.

The rapid adoption of Web GIS has seen it increasingly being used to perform and deliver advanced geospatial analysis to meet a variety of needs. For example, in a chemical spill, dispersion modeling is used to calculate the area affected by the spilled chemicals and overlay analysis is used to calculate the population to be evacuated. In retail, site selection is used to determine the most profitable location to open a new business outlet. In other uses, solar radiation modeling is used to find the best location to place solar panels, and earthquake models can help predict a tsunami. Crime analysis is used to find danger zones that need more police patrols. Optimal routing analysis can be used in a variety of applications to account not only for distance and speed limits, but also time windows, traffic conditions, bridge height barriers, and legal constraints. Over the past half century since the birth of GIS, a vast range of spatial analysis functions has emerged. As GIS is implemented more and more on a Web platform, its power to solve real-world problems will enable organizations to better serve their employees, their customers, and the public.

## 1.4.2 USES

Web GIS functions can be used in a variety of industries, as well as in daily life. Web GIS can reduce costs while bringing increased productivity and efficiency on many fronts. This section includes brief descriptions of several types of applications. You will see more examples throughout the book.

### WEB GIS AS A NEW BUSINESS MODEL AND A NEW TYPE OF COMMODITY

Web GIS has created new business models and reshaped many existing ones. In the process, it has generated tremendous revenue directly and indirectly (see chapter 8).

The most notable new business model is the placement of advertising based on Web mapping. It is the model used by Google, Microsoft, and Yahoo. These Web sites display sponsors' products and services according to the keywords and locations Web users search for. Such placement provides better precision in marketing and a higher rate of return than traditional advertisements. The pay-per-click price model gives sponsors better control over how much to spend on advertising as well as a better understanding of how effective their advertising is. It's not a secret that this business model generates great wealth for the companies that use it.

Web GIS can also be provided as a commodity in itself vis-à-vis the SaaS business model. For example, ESRI Business Analyst Online (BAO) is a Web-based solution (see chapter 8) that combines GIS technology with extensive demographic, consumer spending, and many other types of business data to deliver business functionality. BAO can be used to analyze trade areas, evaluate business sites, and identify the most profitable customers. The service provider charges users by a per-use or subscription price model. Users, who don't have in-house GIS hardware, software, a database, or GIS professionals on hand, can still access the power of GIS through a few mouse clicks and gain sound support in making business decisions.

Many enterprises use Web GIS for strategic planning, marketing, customer service, and daily operations, improving efficiency and gaining competitive advantage. Where once desktop GIS was largely employed for such tasks, it is now Web GIS that is increasingly being adopted. This is because Web GIS is easy to use and can be directly accessed by executives, general employees, regional offices, field employees, and sometimes customers. Almost all business Web sites have functions such as "store locator" that help customers find a store quickly and get directions. A water supply company can integrate Web GIS with its CRM (customer relationship management) database, map customers who complained about low water pressure, identify which valve may be

**Figure 1.11** The up-to-date natural hazards Web map application by USGS keeps citizens informed of earthquakes (red dots), wildfires (yellow icons), flood warnings (green shading), storm warnings (purple shading), and other types of hazards.

Courtesy of U.S. Geological Survey.

failing, trace what other valves need to be shut down, assign the task and necessary maps by mobile device to the nearest field crew, and have the complaints resolved promptly. The use of Web GIS allows the utility to gain user satisfaction. FedEx uses Web GIS to track its vehicles in real time and monitor conditions that affect products such as perishables, which demand a stable temperature range, so that it can ensure timely delivery as well as help customers track their purchases (Mollenkopf 2009).

## WEB GIS AS AN ENGAGING AND POWERFUL TOOL FOR E-GOVERNMENT

E-government and Web GIS have grown up like twins since 1993. Many countries have actively promoted the development of e-government through legislation, regulation, and financial incentives. Geography provides the underlying framework for government activities, and government affairs are generally associated with location. This makes Web GIS a necessary component of e-government (figure 1.11). With easy-to-understand online maps, Web GIS can serve as an engaging communication channel. With its analytical power, Web GIS can deliver broad geospatial intelligence to decision makers (see chapter 9).

Web GIS is a huge facilitator of public services—it keeps citizens informed and helps to create transparency in government. For example, USGS provides current natural hazards maps to alert citizens of earthquakes, hurricanes, and wildfires in near real time and to warn citizens of upcoming gales, storms, and floods. The State of Maryland created MD iMap to provide an overarching look into the performance of Maryland state government. The Web site gives citizens, government staff, and other stakeholders access to a variety of information, such as designated open space, protected ecological areas, and the progress of highway beautification projects (see chapter 9). In California, under Megan's Law, state and county governments provide the locations of sex offenders to help protect the public. Web GIS has proven to be the most effective way to accomplish this goal.

Web GIS can also serve as a bottom-up channel to funnel information from the public to government agencies. For example, the BLM and U.S. Forest Service have experimented in using Web GIS to allow the public to mark or sketch on a Web map whether they support or object to a planned land use to improve the quality of government planning. The U.S. Centers for Disease Control and Prevention provides Web applications to allow the public to report where they see a dead bird, which could be infected by West Nile virus. The addresses collected are mapped so that authorities can collect and examine the dead birds. This process banks on citizen participation to monitor the spread of the deadly disease.

Many government agencies that have long used GIS in their operations are now moving toward the use of Web GIS to take advantage of its strengths in facilitating communication and collaboration (see chapter 9).

## A NEW INFRASTRUCTURE FOR E-SCIENCE

The term e-science refers to science that involves intensive computation or uses immense datasets that require a highly distributed network such as grid computing. Grid computing, which uses a large number of clustered computers to solve a single problem, is usually recognized as the main infrastructure for e-science. Yet the use of grid computing is limited because it requires complex middleware and is not easily accessible by the majority of researchers (see chapter 10). Web GIS promises a new, low-cost alternative with established, readily accessible infrastructure that provides powerful computing capabilities. Its rich datasets will promote advances in e-science.

With the evolution of Web 2.0, the Web has become a distributed database, expanded computing platform, and collaborative lab. There has been an ongoing increase in the number of sensors directly connected to the Web as well as additional rich and real-time datasets. As large enterprise organizations have provided geospatial computing in the cloud through their Web services, and small organizations and the general public have also added geospatial services to the Web, there is now a wealth of powerful analysis capabilities available to scientists over the Web. Scientists can assemble the resources they need through Web programming interfaces without having to be highly trained specialists in grid computing. With the Web as an infrastructure for e-science, the entry costs to working digitally have been lowered, and the benefits are increasingly evident (Hall, De Roure, and Shadbolt 2008).

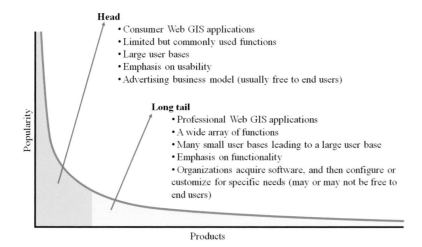

**Figure 1.12** The long tail theory reveals the different market and business models in Web GIS.

### WEB GIS AS AN ESSENTIAL COMPONENT OF DAILY LIFE

As one of the proverbial "five Ws and one H"—who, what, when, where, why, and how—"where" is an essential element of day-to-day existence. Every day you need to answer questions such as where to eat, where to stay, where to shop, and how to get from here to there. It is even important to recognize spatial literacy as a crucial intelligence that should be taught in school so that children grow up as functioning adults in today's world. Spatial literacy can be considered as necessary as reading, writing, and arithmetic—essentially, the "fourth R" (Goodchild 2006). Many people resort to Web mapping sites to acquire such information at the click of a mouse. With the popularity of cell phones and the mobile Web, Web GIS accompanies you wherever you go, and you can obtain location-based services (LBS) whenever you need them (see chapter 5).

### 1.4.3 OPPORTUNITIES

A good way to assess Web-based business opportunities and operation models is by using the long tail theory (Anderson 2004, 2006). According to the long tail theory (figure 1.12), there are as many users at the head, or the top of the curve (the mass market), as there are spread out along the "long tail" of the curve (niche markets). Using music stores and book stores as examples, some stores sell only the most popular record or book titles (the head), while others such as Amazon offer a much wider selection (along the tail). Amazon's success lies partially in selling unique items, even in relatively small quantities, yet the total sales end up being a large number.

To some extent, the long tail theory can also be applied to Web GIS (Bouwman 2005):

- **The head:** Consumer Web mapping applications such as those offered by Google, Microsoft, Yahoo, and MapQuest target the mass market. They have limited but commonly used functions such as mapping, searching for points of interest, finding places and addresses, and routing. Because of their large user base, they can offer the services free to end users and make a profit from their advertising sponsors.

- **The tail:** Many governments, businesses, and institutes have specialized needs that cannot be met with the capabilities offered by consumer Web mapping applications. These needs include collecting unique or private geospatial data, delivering business-specific functions to customers or employees, communicating within the organization and across organizations, and supporting business operations and decision making.

  These specific and diversified needs demand specific solutions. The solutions usually emanate from the organizations themselves, with the help of professional GIS software vendors and GIS consulting firms. Professional GIS software can be configured and customized at the server side and at the client side.

- **Potential markets versus realized markets:** The markets at the head and at the tail of Web GIS are still not fully realized. There is great potential at both ends. Web GIS is still in the early stages. There are scores of niches still untapped and large potential yet to be met. Web GIS offers a whole frontier waiting to be explored.

- **The opportunities:** Consumer Web mapping applications have created a more spatially aware society, exposing geospatial visualization tools to mass users and pointing out the value and broad applicability of geospatial visualization. As the public becomes more familiar with spatial visualization tools and their appetite for geographic knowledge grows, they are beginning to ask more intelligent questions about the world around them and wanting to apply Web GIS to solve wide-ranging issues (Dangermond 2009c). This newfound awareness will generate new specialized needs and niches that fall on the long-tail side, presenting great opportunities for professional Web GIS.

GIS professionals will play an important role in truly realizing the potential of Web GIS. Their mission is to provide authoritative content, serve high-quality visualization tools, deliver analytic models according to specific needs, share geographic knowledge, construct large libraries of services for GIS users inside and outside an organization, and make GIS an integral part of enterprise information systems. For GIS educators, today's students are tomorrow's decision makers and GIS users. Developing spatial literacy is crucial to good decision making that affects global health and the life of the community. Web GIS, as a new, convenient, and fun means of education, can teach students to look at issues from a local to a global scale. It will give them the ability to think spatially and to solve real-world problems (Kerski 2007).

The nineteenth century was the century of railways, the twentieth century the century of highways, and the twenty-first century is the century of the information superhighway. Web GIS continues to speed along this highway as the Web continues to expand the importance of GIS. As GIS becomes an ever more pervasive part of human activity, it is taking the world to a new dimension.

## Study questions

1. What are the Internet, the Web, and the mobile Web? What are the relationships among the three?

2. Define GIS, and explain how it relates to you in your personal life or your profession.

3. When did Web GIS emerge? List a few examples of early Web GIS.

4. What are the principles of Web 2.0? How are they reflected in the latest Web GIS technology?

5. What is Web GIS? How does it relate to distributed GIS and Internet GIS?

6. What are the characteristics of Web GIS?

7. What functions can Web GIS perform?

8. Do you use Web GIS in your work or daily life? What functions do you use? What advantages does it provide? What additional functions would you like to have?

9. Explain the long tail theory in the context of Web GIS. What role do you occupy in Web GIS, and how can you seize the opportunity?

## References

Anderson, Chris. 2004. The long tail. *Wired.* http://www.wired.com/wired/archive/12.10/tail.html (accessed June 30, 2009).

———. 2006. *The long tail: Why the future of business is selling less of more.* New York: Hyperion.

Bouwman, Dave. 2005. *The GIS longtail—Google, MSN, Yahoo, and ESRI.* http://blog.davebouwman.com/?p=853 (accessed June 30, 2009).

CERN. 2008. The Web site of the world's first-ever Web server. http://info.cern.ch (accessed June 17, 2009).

Dangermond, Jack. 2009a. GIS, design, and evolving technology. *ArcNews* (Fall). http://www.esri.com/news/arcnews/fall09articles/gis-design-and.html (accessed January 19, 2010).

———. 2009b. Geospatially enabling information: GIS, the Web, and better government. *ArcUser* (Fall). http://www.esri.com/news/arcuser/1009/geoweb20.html (accessed March 1, 2010).

———. 2009c. GIS professionals lead the GeoWeb revolution. *ArcNews* (Fall). http://www.esri.com/news/arcnews/fall09articles/gis-professionals.html (accessed January 19, 2010).

DiNucci, Darcy. 1999. Fragmented future. *Print* 53 (4): 32.

Douglas, Jane. 2008. 15 years of the World Wide Web. *MSN Tech and Gadgets.* http://tech.uk.msn.com/features/article.aspx?cp-documentid=8197321 (accessed June 30, 2009).

ESRI. 2009. ESRI International User Conference questions and answers. http://events.esri.com/uc/QandA/index.cfm?fuseaction=printall&ConferenceID=2A8E2713-1422-2418-7F20BB7C186B5B83 (accessed November 1, 2009).

Frew, James, Larry Carver, Christoph Fischer, Michael Goodchild, Mary Larsgaard, Terence Smith, and Qi Zheng. 1995. The Alexandria rapid prototype: Building a digital library for spatial information. ESRI International User Conference, Palm Springs, Calif., May 22–26. http://proceedings.esri.com/library/userconf/proc95/to300/p255.html (accessed June 22, 2009).

Goodchild, Michael F. 2006. The fourth R? Rethinking GIS education. *ArcNews* (Fall).

———. 2007. Citizens as sensors: The world of volunteered geography. *GeoJournal* 69 (4): 211–21.

Haklay, Muki, Alex Singleton, and Chris Parker. 2008. Web mapping 2.0: The neogeography of the GeoWeb. *Geography Compass* 2 (6): 2011–39.

Hall, Wendy, David De Roure, and Nigel Shadbolt. 2008. The evolution of the Web and implications for e-research. *The Royal Society A.* http://rsta.royalsocietypublishing.org/content/367/1890/991.full (accessed June 30, 2009).

Huse, Susan. 1995. GRASSLinks: A new model for spatial information access in environmental planning. Unpublished doctoral dissertation, University of California, Berkeley.

International Telecommunication Union. 2009. New ITU ICT Development Index compares 154 countries. `http://www.itu.int/newsroom/press_releases/2009/07.html` (accessed May 27, 2009).

———. 2010. World Internet users and population stats. `http://www.internetworldstats.com/stats.htm` (accessed January 25, 2010).

Internet Society. 2003. A brief history of the Internet. `http://www.isoc.org/internet/history/brief.shtml` (accessed June 23, 2009).

Kerski, Joseph J. 2007. Chapter 6: The world at the student's fingertips: Internet-based GIS education opportunities. In *Digital geography: Geospatial technologies in the social studies classroom,* ed. J. Milson, and Andrew and Marsha Alibrandi. Charlotte, N.C.: Information Age Publishing.

Longley, Paul A., Michael F. Goodchild, David J. Maguire, and David W. Rhind. 2005. *Geographic information systems and science.* 2nd ed. San Francisco: John Wiley & Sons.

Maguire, David. J. 2008. GeoWeb 2.0 and its implications for geographic information science and technology. Proceedings of the Geospatial Information and Technology Association Conference, Seattle, Wash., March 9–12.

Mollenkopf, Adam. 2009. GIS at FedEx. ESRI International User Conference, San Diego, Calif., July 13–17. `http://www.esri.com/industries/logistics/popup_fedex.html` (accessed January 19, 2010).

Nebert, Douglas D. 1995. Serving digital map information through the World Wide Web and wide-area information server technology. Reston, Va.: U.S. Geological Survey.

Netcraft. 2010. January 2010 Web Server Survey. `http://news.netcraft.com` (accessed January 25, 2010).

O'Reilly, Tim. 2005. What is Web 2.0. O'Reilly Network, Sept. 30. `http://www.oreillynet.com/pub/a/oreilly/tim/news/2005/09/30/what-is-web-20.html` (accessed June 28, 2009).

Peng, Zhong-Ren, and Ming-Hsiang Tsou. 2003. Internet GIS: Distributed geographic information services for the Internet and wireless networks. Hoboken, N.J.: John Wiley & Sons.

Putz, Steve. 1994. Interactive information services using World Wide Web Hypertext. First international conference on the World Wide Web, Geneva, Switz., May 25–27. `http://www2.parc.com/istl/projects/www94/iisuwwwh.html` (accessed June 25, 2009).

Scharl, Arno, and Klaus Tochtermann, eds. 2007. *The geospatial Web: How geobrowsers, social software, and the Web 2.0 are shaping the network society.* London: Springer.

Smith, Terence R. 1996. A brief update on the Alexandria Digital Library Project. `http://www.dlib.org/dlib/march96/briefings/smith/03smith.html` (accessed June 26, 2009).

Steinitz, Carl. 1990. A framework for theory applicable to the education of landscape architects (and other environmental design professionals). *Landscape Journal* 9 (2): 136–43.

Tomlinson, Roger F. 2008. *Thinking about GIS: Geographic information system planning for managers.* 3rd ed. Redlands, Calif.: ESRI Press.

Turner, Andrew. 2006. *Introduction to neogeography.* Sebastopol, Calif.: O'Reilly Media.

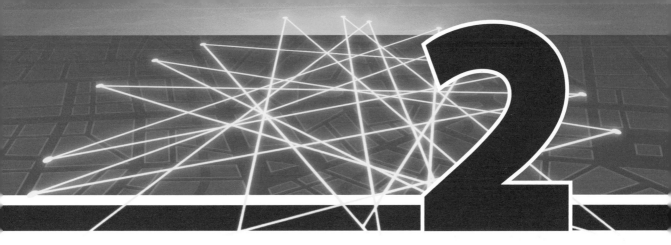

# TECHNICAL BASICS

Web GIS is an exciting new area of application development. You probably have used various types of Web GIS applications: finding pizza restaurants near your home, finding business locations near your hotel, or tracking delivery fleets at work.

This chapter covers the basic concepts and technology behind Web GIS applications. Section 2.1 presents the basic technologies of the Web, including its three principal technologies—HTTP, URL, and HTML—as well as related technologies on the server side, the browser side, and the data transmission between them. Section 2.2 introduces the major components of Web GIS, which include the Web GIS server, the GIS database, and the client. Geobrowsers and the online virtual globe are defined in this section as popular and ready-to-use Web GIS clients. Section 2.3 talks about the thin client and thick client architectures, and then introduces a design pattern that is applicable to most Web GIS applications. Section 2.4 highlights design based on the user experience. While not intended to be a step-by-step guide or a "how to," the contents of this chapter can familiarize you with the basic technologies and principles of Web GIS applications design.

## 2.1 WEB FUNDAMENTALS

Web application architecture can be complex, because one application can be interconnected with other applications. This type of architecture will be explained later in the book. This section introduces the basic architecture of a Web application.

### 2.1.1 WEB PRINCIPLES

**A basic Web application is essentially a client/server (C/S) architecture, in which the client is specifically called a Web client, and the server is called a Web server.** Basic Web applications usually have a three-tier architecture consisting of the data tier, logical tier, and presentation tier (figure 2.1). The Web client is typically a Web browser—thus, the basic Web architecture is also referred to as the browser/server (B/S) architecture. The basic workflow of a Web application is as follows:

- A user uses a Web client, usually a browser, and initiates a request to the Web server by typing a URL in the address bar or clicking a URL link on a Web page.

- The Web server receives the request, parses the URL, locates the corresponding document or script, and returns the document or else executes scripts and returns the result of the script to the client as a response. The response is commonly in HTML format.

- The Web client (the browser) receives the response, renders it, and presents it to the user.

**The three foundational linchpins of the World Wide Web (WWW) are HTTP, the URL, and HTML.** These protocols were invented by Tim Berners-Lee in 1990. These technologies are now international standards, and their specifications are managed and maintained by W3C (World Wide Web Consortium).

#### HTTP

Because it involves multiple parties, the Web needs a protocol that defines the rules and procedures for all parties to follow. As a protocol, **HTTP (Hypertext Transfer Protocol) defines a set of rules**

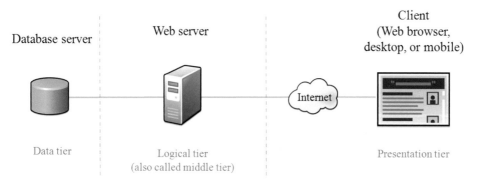

**Figure 2.1**   The basic architecture of a Web application includes three tiers.

**and procedures that both Web clients and Web servers use to communicate with each other.** For example, with HTTP, a Web server knows what information to put in the message header and what to put in the message body, and the Web client knows what to expect from both the response header and body.

HTTP defines eight methods of request: GET, POST, HEAD, PUT, DELETE, TRACE, OPTIONS, and CONNECT. GET and POST are the two most commonly used requests. Both HTTP request and response messages have headers and bodies. The following information is typical of what can be found in an HTTP response header:

- **Cache control:** specifies directives that the client must obey (e.g., no cache)

- **Content type:** the MIME (Multipurpose Internet Mail Extensions) type indicates whether the message body is HTML (text/.html), a JPEG image (image/.jpeg), an MP3 audio file (audio/.mpeg3), or other file type

- **Status code:** denotes the status of the response (e.g., 200 means success, 403 means forbidden, and 500 means server error)

HTTP is characterized as simple, stateless, and flexible:

- **Simple:** In general, an HTTP communication scenario has three steps: (1) an HTTP client makes a connection to a Web server and sends a request method to the HTTP server; (2) the server processes the client's request while the client waits for the server's response; and (3) the server responds with the status code in the response header and data (if any) to the client and closes the connection.

- **Stateless:** After the server has responded to a client's request, the connection between the client and the server is typically dropped and forgotten. Servers do not retain information about the client between requests, which can greatly reduce the load on the server.

- **Flexible:** HTTP allows the transmission of any type of data object, and the content type can be specified in the HTTP header.

**HTTPS (Secure Hypertext Transfer Protocol) refers to the use of ordinary HTTP over an encrypted Secure Sockets Layer (SSL) connection.** With a plain HTTP connection, the data transferred between the client and the server can be intercepted. HTTPS avoids eavesdropping by encrypting the data. This ensures reasonable protection over the Internet, which is largely an insecure network. HTTPS is often used to transfer sensitive data such as personal user profiles, login passwords, and credit card information. Supporting HTTPS on a Web site requires installation of a certificate on the Web server that is issued by certification authorities.

## URL

**A URL (Uniform Resource Locator) is a subset of the URI (Uniform Resource Identifier) that specifies where an identified resource is available on the Internet and the mechanism for retrieving it.** A URL is often called a Web address. Every Web page has a globally unique URL. Just as a street address is used to locate a household in the real world, a Web address is used to locate a Web page on the World Wide Web, which holds billions of Web pages. Without a URL, Web clients would not be able to reach a Web server or the desired resources on the Web server. Without URLs, the Web resources could not be interlinked to form the Web.

The basic syntax of a URL is

```
Protocol: //hostname[: port]/filepathname?query_string#anchor
```

- Protocol specifies the transport protocol between the client and the server. Typical examples include HTTP, HTTPS, FTP, and MMS.

- The host name or IP address specifies the Web server to be reached. Sometimes, before the host name, there is a user name and a password, which can be used to connect to the server. The format is "username@password."

- The port number is optional. When the option is omitted, the default port for the protocol is used. For example, the default port for HTTP is 80, and for HTTPS is 443.

- The file path indicates the directory and file name of a resource on the Web server. Sometimes, the file path is an alias instead of a real directory or a real file name.

- The query string is optional and is one way to pass parameters to a script on the Web server. The query string normally contains one or more name-value pairs separated by ampersands, with names and values in each pair separated by an equal sign.

- The anchor in the URL specifies a location or a section on a Web page.

### HTML

**HTML (Hypertext Markup Language) is the main language for creating Web pages and is the source code for most Web pages.** Similar to a Word document, HTML contains contents, a layout, and formatting information. When a Web page is loaded into a Web browser, the Web browser interprets the HTML code and displays the Web page the way the Web designer intended it to look. As a markup language, HTML is essentially plain text marked by a set of tags, such as head, body, table, center, and fonts. The appearance and layout information for HTML can be defined in Cascading Style Sheets (CSS), a scripting language used to style Web pages. CSS can be directly embedded in HTML files, or be in a separate file that is referenced in HTML files.

Web pages built with pure HTML usually have a plain user interface. HTML allows images, scripts, and objects to be embedded. Scripts such as JavaScript and objects such as browser plug-ins can implement logic and workflows to enhance the user interface.

### 2.1.2 RELATED TECHNOLOGY

This section covers Web application development technologies for the Web server and the Web client and the communications between them (figure 2.2).

### SERVER-SIDE TECHNOLOGY

Server-side technology includes Web application servers and a range of programming languages that run inside them.

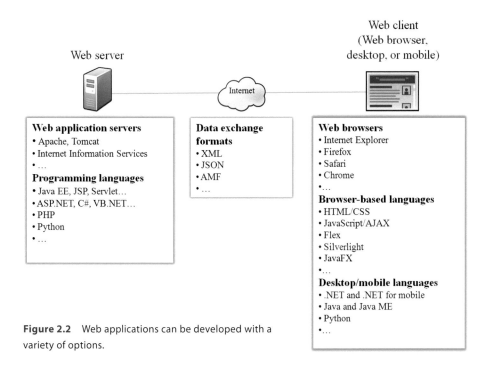

**Figure 2.2** Web applications can be developed with a variety of options.

## WEB APPLICATION SERVER

Web application servers host the Web sites that you read and serve the data for other Web applications to use. They are responsible for accepting HTTP requests from clients and serving clients with HTTP responses. Since they comply with HTTP specifications, Web application servers know how to communicate with Web clients. Web application servers are also containers that allow certain scripts such as JSP (JavaServer Pages), Java Servlet, and ASP.NET to run inside. Such scripts allow Web developers to implement certain business logic.

Examples of Web application servers include the following:

- **Apache Web Server and Tomcat** from the Apache Software Foundation: Apache is an open-source Web server characterized by simple, fast, and stable performance. It can run on almost all UNIX, Windows, and Linux system platforms. It is the most widely used Web server software (Netcraft 2009) and can support the Java suite of programming languages, and other languages when proper extensions are installed.

- **IIS** (Internet Information Services, formerly Internet Information Server), from Microsoft: IIS is a Web server application for use with Microsoft Windows. It is the world's second most popular Web server (Netcraft 2009) and can support the .NET suite of programming languages, and other languages when proper extensions are installed.

- Other examples of Web application server products include Nginx, Oracle/Sun Java System Web Server, and IBM WebSphere Web Application Server.

*SERVER-SIDE PROGRAMMING TECHNOLOGY*

Server-side programs execute on the Web server and within the Web application server container. The main programming languages include

- **JavaEE (Java Enterprise Edition), Servlet, JSP, and JSF (JavaServer Faces),** technologies of the Java family: These technologies are widely used for developing Web applications. James Gosling, coinventor of Java, once joked that "Java is C++ without the guns, knives, and clubs" (1996). Java is similar to the C++ programming language, though it drops some C++ features that are tricky if not used right. Web applications developed with Java can run inside Apache, Tomcat, the Oracle/Sun Java System Web Server, and IBM WebSphere Web Application Server.

- **ASP.NET,** an important part of the Microsoft .NET framework: The .NET framework is a common environment for building, deploying, and running Web services and Web applications. The .NET framework has technologies for both server-side and client/browser-side programming development. ASP.NET has a set of libraries that facilitate users writing ASP.NET Web pages (.aspx) in C# (a language that is different from C++ and Java though built with similar syntax and semantics to C++ and Java) and VB.NET.

## CLIENT/BROWSER-SIDE TECHNOLOGY

Web clients include a variety of programs and applications that can run inside or outside the browser, including desktop applications and mobile applications, which run outside the browser. In this section, we focus on the Web browser, the most popular type of Web client.

*WEB BROWSER*

A Web browser is a software application for retrieving and presenting information resources on the Web. It provides a means for looking at and interacting with the Web. For most users, Web browsers represent the "face" of the Web. Technically, a Web browser is a client that implements HTTP specifications, HTML specifications, and JavaScript specifications, meaning that it knows how to communicate with Web servers, how to display an HTML, and how to interpret and execute JavaScript.

The most popular Web browsers are Internet Explorer (IE) and Mozilla Firefox (Net Applications 2009). Other Web browsers include Apple Safari and Google Chrome. There are slight differences in how these browsers support HTML and JavaScript specifications, meaning that the same Web page may appear and behave differently in different Web browsers.

*BROWSER-SIDE PROGRAMMING TECHNOLOGY*

- **JavaScript,** invented by Netscape in 1995, is a scripting language that primarily runs inside Web browsers to make Web pages dynamic and interactive. It is the most popular scripting language, used in millions and millions of Web pages. JavaScript is not Java, although its basic syntax is intentionally similar to both Java and C++ to reduce the number of new concepts required to learn the programming language. JavaScript is simple so that nonprofessional programmers can work with it, and it is safe to use because it has limited access to the local hard drive of the client computer where it is running. JavaScript is usually platform independent, but there are circumstances when a JavaScript behaves differently in different browsers. Developers who want to build cross-browser applications need to be careful about possible incompatibility. The combination of HTML and JavaScript is called DHTML (dynamic HTML) and can be used to create Web sites that are interactive as opposed to static.

- **AJAX (Asynchronous JavaScript and XML)** has become popular since 2005. It is not a new programming language but a group of existing Web development techniques that are used to create better, faster, and more interactive Web applications on the browser. AJAX uses asynchronous communication between the browser and the Web server so that a Web page can retrieve data from the server in the background (asynchronously) without interfering with the display and behavior of the existing page. Without AJAX, you generally have to move from page to page and wait for a new page to be loaded before you can interact with it. With AJAX, you can remain on a page while the data needed from the server is retrieved, in XML or other formats, and then parsed and updated on the same page. The main contribution of AJAX is improvement of the user experience. From a software engineering point of view, AJAX allows developers to separate data from formatting, which is the preferred software design pattern.

- **Adobe Flex** is an open-source framework for building highly interactive, expressive Web applications that deploy consistently in all major browsers. It can present sophisticated animation and transmission effects, which makes it a powerful language for implementing rich and engaging Web applications, or rich Internet applications (RIAs, see section 2.4.2). Flex uses MXML, an XML-based language, to describe user interface layouts and behaviors, and ActionScript, a strong-type, object-oriented programming language, to create client logic. Programs developed with Flex can run inside Web browsers with the Adobe Flash Player plug-in or outside Web browsers on Adobe AIR, the cross-operating system runtime environment. This enables Flex applications to run consistently across all major browsers and platforms.

- **Microsoft Silverlight** is a cross-browser and cross-platform plug-in for delivering RIA that uses graphics, animation, or video within the .NET framework. It uses XAML (Extensible Application Markup Language) to develop the user interface, and .NET programming languages such as C# or VB.NET to develop business logic. Microsoft Silverlight programs can run inside Web browsers with a Microsoft Silverlight plug-in or on the desktop with a WPF (Windows Presentation Foundation) runtime environment. This makes Microsoft Silverlight a common method for developing browser-based and desktop applications.

- **Oracle/Sun JavaFX** is the Oracle/Sun software platform intended for creating and delivering RIAs that can run across a wide variety of devices. JavaFX is fully integrated with the Java Runtime Environment (JRE), which ensures that JavaFX applications can run in any browser, desktop, or mobile phone with the proper JRE installed.

## DATA EXCHANGE FORMAT

Contemporary Web technology prefers to separate the data from the formatting and the exchanging of data from the presentation. The data format used for exchanging messages between the server and the client can greatly influence the load to bandwidth and therefore the efficiency of the application. There are three main formats for exchanging data over the Web:

- **XML (Extensible Markup Language)** is a markup language that allows you to define your own tags and attributes. XML has many strengths in that it is platform independent because it is a plain text file; it is self-descriptive since its tags and attributes usually have easy-to-understand names; it can be processed automatically, because it is well structured; and it can be validated against its schema. Because of these advantages, XML is probably the most commonly used data exchange format on the Web. However, it also has some disadvantages. It is bulky because it uses tags to delimit the data. And parsing XML (i.e., extracting values from certain nodes in an XML) is not efficient, especially in languages such as JavaScript, the most commonly used scripting language on the Web.

- **JSON (JavaScript Object Notation)** is a lightweight computer data interchange format. It is derived from the object literals of JavaScript, as defined in the ECMA (European Computer Manufacturers Association) Script Programming Language Standard. JSON is smaller in size than XML, and it is more efficient to parse in JavaScript where JSON is a native data type. The main application of JSON is in Web application programming, where it serves as an alternative to the XML format. The following example uses XML and JSON to describe the same information concerning the names and hobbies of two students.

```
An XML example
<?xml version="1.0" encoding="UTF-8"?>
<students>
    <student>
            <name>John</name>
            <hobby>Basket Ball</hobby>
    </student>
    <student>
            <name>Lisa</name>
            <hobby>Movie</hobby>
    </student>
</ students >

A JSON example

{"students": [
            {"name":"John", "hobby":"Basket Ball"},
            {"name":"Lisa", "hobby":"Movie"}
        ]
}
```

- **AMF (Action Message Format),** developed by Adobe, is a binary format used primarily for exchanging data between Adobe Flex/Flash Web client applications and Web servers. AMF is a native data type in Flex, making it more efficient to process than JSON. Compared with XML, which is a native data type in Flex as well, AMF is much smaller than XML, making AMF more efficient to transfer and process. Although AMF is mostly used in Flex applications, there are implementations of AMF in PHP, Java, and .NET as well.

## 2.2 WEB GIS BASIC ARCHITECTURE AND COMPONENTS

**Web GIS extends a basic Web application by giving it GIS capabilities.** The basic architecture of Web GIS is similar to Web applications but with the addition of GIS components (figure 2.3). In a basic workflow, a user uses a Web GIS application through a client, which can be a Web browser, a desktop application, or a mobile application. The client sends a request to the Web server over the Internet via HTTP. The Web server forwards GIS-related requests to the GIS server. The GIS server retrieves the needed data from the GIS database and processes the request, which can be to generate a map, conduct a query, or perform an analysis. The data, map, or other result is sent by the Web server to the client in a response via HTTP. The client then displays the result to the user, which completes the request and response cycle.

### 2.2.1 WEB GIS SERVER

**The Web GIS server is the most important component in a Web GIS. Its functionality, ability to be customized, scalability, and performance are critical to the success of the Web GIS application. The capability and quality of a Web GIS application is largely determined by the Web GIS server it uses.**

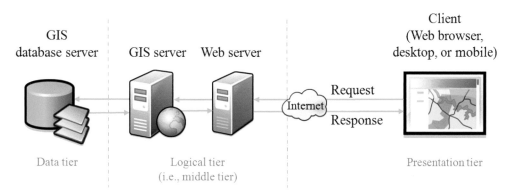

**Figure 2.3**    In the logical architecture and workflow of a basic Web GIS, the components can all be deployed on one computer, but logically, they are still three separate tiers.

GIS server technology has been evolving over the last two decades. Using ESRI products as an example, the first-generation commercial GIS server products ArcView IMS and MapObjects IMS were released around 1996. They represented an important addition to the GIS software family and revealed a new direction in GIS software products moving toward the Web platform. The second-generation commercial GIS Server released by ESRI was ArcIMS, around 1998. It offers enhanced performance and extended functionality, but it lacks the extensive functionality of the third-generation product, ArcGIS Server. ArcGIS Server offers a more comprehensive set of functionality (figure 2.4), including the following:

- Dissemination of 2D maps, 3D globes, query, search, feature editing, data extraction, tracking, imagery, address and place finding, routing, geometry processing (e.g., transferring coordinate system), and metadata capabilities to the Internet or intranet.

- Support of advanced and customizable geoprocessing capabilities.

- Delivery of capabilities via Web services, which are Web-based programming components that can be reused and remixed to create new applications. Web services technology can shrink data and application redundancy, optimize system configurations, and consolidate enterprise systems. The recent popular REST (Representational State Transfer)-style Web services have introduced many new benefits in addition to the traditional SOAP (Simple Object Access Protocol) Web services (see chapter 3).

- Compliance with industry standards, such as OGC (Open Geospatial Consortium)-compliant WMS, WFS, WCS, CSW, GML, and KML (see chapter 3). This enables Web-service-level interoperability (e.g., plug and play) among products by different vendors.

- Ways to achieve high-quality services, including cached services, optimized services, and other methods to accomplish fast performance and high scalability.

- Token services and other methods to secure geospatial Web services (see chapter 3).

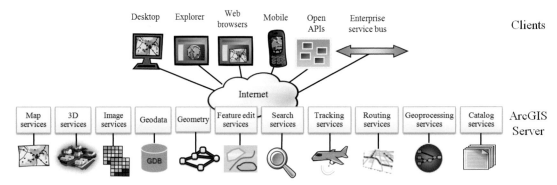

**Figure 2.4** ArcGIS Server provides an array of functions to a variety of clients via Web services.

- Ability to be highly customized. Supports a variety of modern client-side and server-side APIs that can build highly interactive and engaging client applications for Web browsers, desktops, and mobile devices (see "Web GIS clients" below).

## 2.2.2 GIS DATABASE

The GIS database is the data storage and management framework for a GIS. It can hold a collection of geographic datasets of various types, such as basic vector data (points, lines, and polygons) and raster data (satellite and aerial images, for instance). Some GIS databases support more data types such as CAD data, 3D data, utility and transportation system data, GPS coordinates, and survey measurements. GIS databases range from a small, single-user database up to a larger, enterprise-level database, which can be edited and accessed by many users simultaneously. While some GIS databases store only collections of individual features, others manage the data models that define the spatial relationships and behaviors that are critical to many GIS tasks and analytical operations.

**The GIS database is the underlying support for Web GIS applications. The answers delivered from a Web GIS application can only be as good as the quality of the information contained in the GIS database. While casual applications can rely on casual data sources, professional applications tend to need high-quality, authoritative, and up-to-date geographic information.**

A strong GIS database product generally includes the following capabilities:

- Stores a rich collection of spatial data in centralized or distributed systems
- Applies sophisticated rules and relationships to the data
- Defines geospatial relational models (e.g., topologies, road networks)
- Maintains integrity of spatial data with a consistent and accurate database
- Works within a multiuser access and editing environment and supports versioning
- Supports custom features and behavior
- Provides a means for robust data security, backup, recovery, and rollback
- Maintains high performance when the volume of data increases and the number of simultaneous users increases.

## 2.2.3 WEB GIS CLIENTS

The client in a Web GIS application can play two roles. First, it represents the end-user interface for the entire system. It interacts with the user, collects user inputs, sends requests to the server, and presents the results to the user. Most end users don't know, and don't need to know, the particulars of the back-end server(s). All that they know about the system is how fast, stable, and user-friendly the client is. Second, the client, especially a thick client (see section 2.3), can also perform some geospatial processing tasks, such as dynamic classification for thematic mapping, cluster, and heatmap analysis. Web GIS clients are typically Web browsers but can also be desktop applications, mobile applications, or even server applications (when a server acts as a client of another server).

## WEB BROWSER CLIENT

Web browsers are the main type of Web GIS client. Early browser clients were static, plain, and dull by today's standards. More recently, Web browsers have surpassed the static HTML and simple JavaScript stage. AJAX, Flex, Silverlight, and JavaFX technologies can create a rich, dynamic, and user-friendly interface known as an RIA (see section 2.4.2). With additional APIs such as ESRI's ArcGIS API for JavaScript, Flex, or Silverlight, Web browsers are now able to easily perform many types of GIS functions (see chapter 4). Examples of Web browser clients include ArcGIS Explorer Online.

## DESKTOP APPLICATION CLIENT

Desktop applications such as those developed with Java, .NET, or other programming languages can act as Web GIS clients. These full-featured applications do not run inside a Web browser, so they are not limited to the "sandbox" environment of a browser (i.e., a tightly controlled set of resources for guest programs to run in). Desktop clients can access local resources, such as local files, the local database, and peripheral devices. They are rigorous and especially fit for resource-intensive applications.

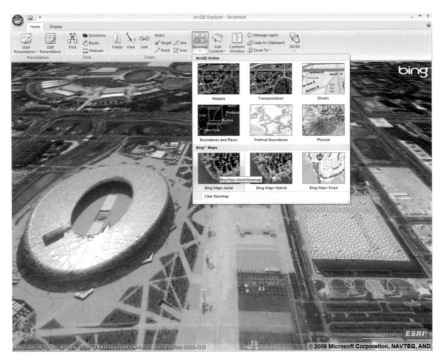

**Figure 2.5** ArcGIS Explorer Desktop is a desktop Web GIS client that can view map services from ArcGIS Online, Microsoft Bing Maps, OGC standard map services, and users' own ArcGIS Server Web services.

Courtesy of Microsoft, NAVTEQ, and AND Automotive Navigation Data.

Examples of desktop clients include ArcGIS Explorer Desktop (figure 2.5), Google Earth, and ArcGIS Desktop. The role of many traditional desktop GIS products has been changing—in addition to running alone or being the client of a local server, they can also be used as the client of Web GIS applications. They can be a good solution when browser-based clients are not powerful enough, especially for professional users. For example, ArcGIS Desktop can use the maps and many other types of services available over the Web, or in the cloud, and perform many advanced operations that are not available in browser clients.

## MOBILE CLIENT

Mobile devices, which allow the use of GIS on the go, are becoming a popular type of Web GIS client. Mobile clients basically can be classified into two categories: browser based and native application based. The capabilities of mobile browsers vary greatly. While some mobile browser-based clients can only deliver simple GIS functions with a plain user interface, others are getting close to being like the fully fledged Web browsers we use on desktops that can deliver a greater variety of GIS functions with a richer user interface. Native mobile applications, developed with .NET Mobile, Java Mobile Edition (Java ME), or other languages, have the added advantage of being able to access local peripheral devices such as GPS to acquire an accurate location of the mobile user and enable a range of applications such as survey, rescue, real-time traffic, navigation, and locating such things as businesses or friends (see chapter 5).

Examples of native mobile application clients (figure 2.6) include ArcGIS Mobile, ArcGIS for iPhone, Microsoft Bing Maps Mobile, Google Earth Mobile, and MapQuest Mobile (see chapter 5).

While most Web GIS clients are designed for specific applications, some clients can work with a variety of Web content types. Such clients are called geobrowsers. **Geobrowsers refer to map viewers that provide the ability to visualize standard Web map services and data available over the Web.** The standards for Web services, data, or maps available over the Web, which aim

**Figure 2.6** ArcGIS Mobile and ArcGIS for iPhone are clients of ArcGIS Server. They can display map services and query data hosted remotely on ArcGIS Server, delivering GIS capabilities to the field. Left: the ArcGIS Mobile out-of-the-box application shows the default user interface. Right: ArcGIS for iPhone shows a basemap overlaid with utility networks.

Courtesy of Tele Atlas North America, Inc., and Trimble.

to achieve interoperability and allow the products of different vendors to work in a "plug and play" environment, are explained in chapter 3. Here are some examples of geobrowsers:

- The Carbon Project Gaia is a 2D geobrowser. It can access an array of geospatial resources such as the OGC WMS, WMTS, WCS, WFS, KML, GML, Microsoft Bing Maps, Yahoo Maps, and OpenStreetMap, as well as file formats such as the ESRI shapefile format.

- ArcGIS Explorer Desktop can be used as either a 2D or 3D geobrowser. It can display data and service standards such as OGC WMS, KML/KMZ, GeoRSS, ArcGIS Server services, and ArcIMS services.

- ArcGIS Desktop is also a geobrowser in that it can view and query OGC WMS, WFS, and WCS, as well as ArcGIS Server services.

A virtual globe is another type of popular Web GIS client. **A virtual globe is a 3D software model or representation of Earth or another world** (Butler 2006; Foresman 2008). It allows you to freely

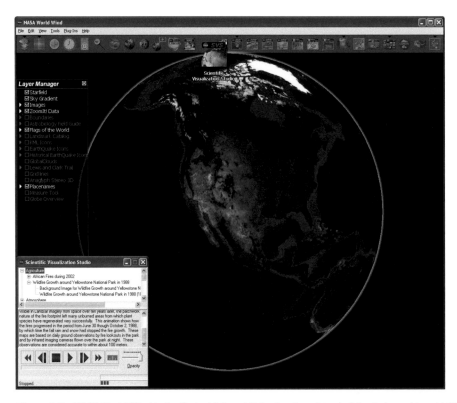

**Figure 2.7** NASA World Wind is the first widely publicized online virtual globe (released in mid-2004). It lays NASA and USGS satellite imagery, aerial photography, topographic maps, and publicly available GIS data on top of 3D models of Earth or other planets.

Courtesy of NASA.

move around in the virtual environment by changing the viewing angle and position. Many virtual globes fetch and display data from the Web and are thus named online virtual globes. **An online virtual globe is a virtual globe that relies on the Web for its data and maps.**

Online virtual globes are mostly free to use. Some popular ones include NASA World Wind (figure 2.7), Google Earth, ArcGIS Explorer Desktop, Microsoft Bing Maps, and SkylineGlobe. Most online virtual globes are also geobrowsers, because they can drape and visualize a variety of standard geospatial contents across their surface.

## 2.2.4 CHALLENGES

A number of bottlenecks and other factors can challenge the quality of a Web GIS application (figure 2.8). They include the following:

- Pressure on the GIS server to support many users
- Pressure on the GIS database to support frequent data read/write
- Pressure on the Internet to frequently transfer large volumes of data
- Limited GIS capabilities of the client, especially Web browsers
- Inexperienced nonprofessional end users

Various solutions are introduced throughout the book, including Web services optimization, such as caching in advance, algorithm and system tuning, failover and load balance, and efficient use of Internet bandwidth (see section 3.5), proper workload partition between the client and the server (section 2.3), and effective user experience design (section 2.4).

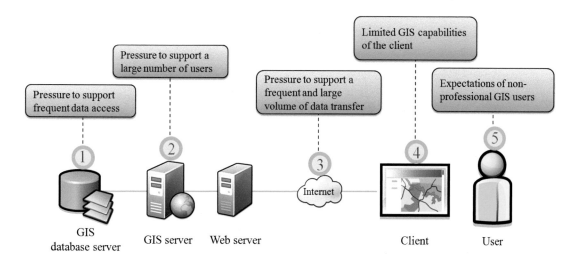

**Figure 2.8** A variety of bottlenecks can challenge the quality of a Web GIS application.

## 2.3 THIN VERSUS THICK CLIENT ARCHITECTURE

A basic design consideration for any client-server system is how to partition the workload between the client and the server. Depending on how the workload is distributed, Web GIS applications can be categorized as either thin client architecture or thick client architecture (Gong 1999).

Both Web GIS servers and clients are becoming more powerful as technology advances, with the trend moving toward thick client architecture. But how thick should the client be? What is the best strategy to split the work between the server and the client? This section presents the best practice for addressing these questions.

### 2.3.1 THIN CLIENT ARCHITECTURE

Thin client architecture relies on the server to perform most of the work, leaving the client to do the least amount. The client simply sends the user's requests to the server. The server does the processing, such as generating a map and performing analysis. The results, typically in HTML format embedded with GIF, PNG, or JPEG images, are then returned to the client and displayed for the user.

An example of thin client architecture is the PARC Map Viewer (figure 1.6), the first Web GIS application in the world. The advantages of this approach are the following:

- The user doesn't need to install any software other than a Web browser, not even a plug-in.
- Since most of the heavy-lifting processes are done by the server, the client doesn't need a powerful computer, so these types of applications can work well even on low-end computers.

The main disadvantages are the following:

- Pressure on the GIS server: all GIS operations are routed to the server and processed by the server.
- Limited user interaction: the user interface is usually built with plain HTML and limited JavaScript, so the user interaction is not particularly smooth or compelling.

### 2.3.2 THICK CLIENT ARCHITECTURE

A thick client is at the other end of the spectrum from the thin client. Thick client architecture relies on the client rather than the server to perform most functions. This is usually accomplished with a Web browser plug-in or a native client application. The client plug-in or native application executes locally on client computers. The thick client requests the source data (e.g., the coordinates of vector data) from the server, and then renders maps and performs analysis on the client side.

The following are typical advantages of this architecture:

- Fast interaction with the user as the logic, or program, runs locally and data resides locally
- Less pressure on the server since there are fewer round trips to the server

The following are the disadvantages:

- Inconvenience associated with the installation of browser plug-ins or installation of native applications. There are cases when the user is not allowed to install plug-ins or native applications —on some government computers or company computers, for example.

- Limitations posed by the Internet bandwidth and client computing power. Typically, it is not feasible to transfer gigabytes of data over the Internet and have the client perform sophisticated GIS operations.

## 2.3.3 BEST PATTERN FOR WORKLOAD PARTITIONING

Web technology advances rapidly. Client-side technology such as plug-ins and JavaScript is becoming more and more powerful and can take on greater and greater workloads. A contemporary popular design strategy involves breaking down the workload into several categories and properly distributing it between the server and the client (figure 2.9).

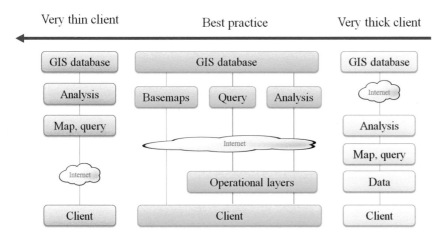

**Figure 2.9** In the extreme thin client architecture, most GIS functions are performed by the server, while the opposite is true for extreme thick client architecture. Current best practice recommends basemaps done by the server and operational layers typically rendered by the browser, unless the data size is too large for the browser to handle. Essentially, easier functions are performed by the browser and complex functions are handled by the server.

This best-practice strategy (equation 2.1) recommends breaking the typical Web GIS application functions into basemaps, operational layers, and tools (Quinn 2008; Brown et al. 2008).

---

**EQUATION 2.1**

Web GIS application = basemaps + operational layers + tools

Where

- basemaps are usually precreated or created dynamically by the server;
- operational layers are typically dynamically rendered by the client, but should be rendered by the server if the data size is too large; and
- easier tasks are done by the client and complex ones by the server.

---

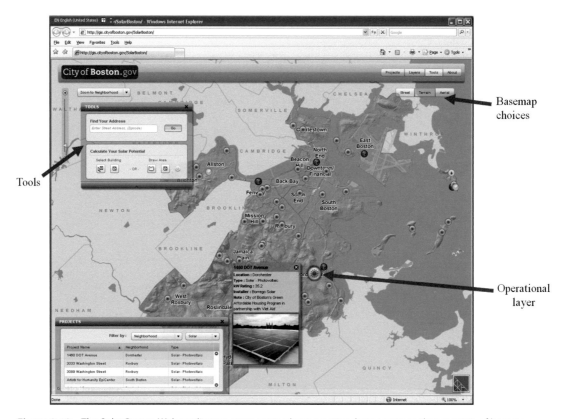

Tools

Basemap choices

Operational layer

**Figure 2.10** The Solar Boston Web application represents a best-practice design pattern that consists of basemaps, operational layers, and tools.

Courtesy of Boston Redevelopment Authority.

The Solar Boston Web GIS application (figure 2.10) is an example of best-practice design. Solar Boston is part of the Solar America Initiative, a campaign launched by the U.S. Department of Energy (DOE) to make solar electricity cost-competitive with traditional electricity production by 2015. To help meet the city's greenhouse gas reduction targets and support the goals of DOE, the City of Boston set a target of 25 megawatts of solar power to be installed by residents and organizations by 2015. To support the Solar Boston program, Boston is using Web GIS technology to map current solar installations, track progress toward the goal, and allow Bostonians to analyze their rooftop solar energy potential (Knight 2009). The Solar Boston Web GIS application was developed by the Boston Redevelopment Authority (BRA) in 2008 using ArcGIS Server and an early version of ArcGIS API for Flex.

## BASEMAPS

Basemaps provide the locational reference or context for an application. They serve as the foundation for all subsequent operations. Best practice recommends that basemaps be generated by the server. Since a basemap tends to be relatively static, and in a typical setting is updated on an infrequent basis, basemaps should be cached, meaning the server should generate a set of map tile images in advance at predetermined scale levels for rapid display (see section 3.5.1). A basemap should be multiresolution and designed to be portrayed appropriately at a range of supported map scales.

The Solar Boston application provides three types of basemaps: a street map, terrain map, and aerial imagery. These basemaps were cached in advance by ArcGIS Server at multiple scales ranging from 1:80,000 to 1:1,250. The maps are based on the City of Boston's authoritative and detailed data, including the rooftops of each building. The street maps and terrain maps allow users to see the streets and building footprints as they zoom in from city level to building level. The terrain maps use the hillshade effect to portray the topography of the city, while the high-resolution aerial imagery gives users a sense of the city's geography. These basemaps provide context for the application and let users get oriented on what Boston looks like, where the green energy projects are located, and how these projects relate to other features such as surrounding buildings.

## OPERATIONAL LAYERS

Operational layers are drawn on top of the basemap and contain the themes that end users will view and work with. Operational layers consist of, but are not limited to, the following types:

- **Observations or sensor feeds.** This material can be information that reflects status or situational awareness—for example, crime locations, traffic sensor feeds, real-time weather, readings from meters (such as stream gauges), observations from equipment or made by workers in the field, inspection results, addresses of customers, disease locations, or air quality and pollution monitors.

- **Editing layers.** These are the map layers that users work with—for example, to edit features. Common examples are the facilities layers edited by a utility worker.

- **Query results.** In many cases, applications will make a query request to the server, which will return a set of records as results. These can include a set of individual features or attribute records. Users often display and work with these results as map graphics in their Web GIS map applications.

- **Result layers that are derived from analytical models.** GIS analysis can be performed to derive new information that can be added as new map layers, and then explored, visualized, interpreted, and compared by end users.

Typically, operational layer data is small in size or is only displayed when users zoom to a certain level. The data is streamed to the client, and then rendered and managed by the client. There are several reasons behind this pattern. First, operational layers are often dynamic, and thus generally should not be cached in advance. Second, users need to interact with these layers to conduct their daily work. They expect operational layers to be responsive, reporting details on a mouse click and showing links to certain functions. The dynamic rendering and user interaction fall on the client side and can be implemented using browser APIs such as ArcGIS APIs for JavaScript, Flex, or Silverlight (see chapter 4). Occasionally, operational layer data can be too large to be effectively rendered by the browser. In this case, operational layer data should be rendered by the server.

In the Solar Boston application, the green energy projects layer is the main operational layer. The coordinates and attributes of the projects are extracted from the server to the browser. The browser uses the ArcGIS API for Flex to dynamically classify each project, displaying different icons for different types of projects (solar, biomass, water, or wind). When the mouse pointer hovers over a project, the project will display its attributes along with a photo in the message window.

## TOOLS

Web GIS applications usually provide tools that perform processes beyond mapping. These tools range from the common types, such as finding an address or place-name, routing, and searching points of interest that match certain criteria, to more specialized tools that implement specific business logic for an enterprise. The solution varies as to whether the processes should be done by the server or the browser.

- **Have the client do it:** This method fits processes that are relatively easy and when the data needed is all on the client side. Typical examples include charting analysis results and generating heat maps based on a set of points (see figure 4.8 B and D).

- **Have the server do it:** This fits processes that are complex and when the data needed is not housed on the client side. Typical examples include finding and routing to the closest facility, calculating stream flows, and finding the best habitat by overlaying a number of data layers.

The Solar Boston application allows you to select a rooftop or draw a polygon on the map, and then calculate your monthly solar energy potential (figure 2.11). The solar potential calculation is computational intensive (Fu and Rich 2002). The calculation needs to be handled by the server because the algorithm is too complex for the browser and because the digital elevation model and building height needed by the algorithm are too large to be streamed to the browser. The calculations are done by the server in advance for existing rooftops (see section 3.5) and are done dynamically for the areas that users draw. The resultant monthly solar energy potentials are sent to the browser and displayed in charts.

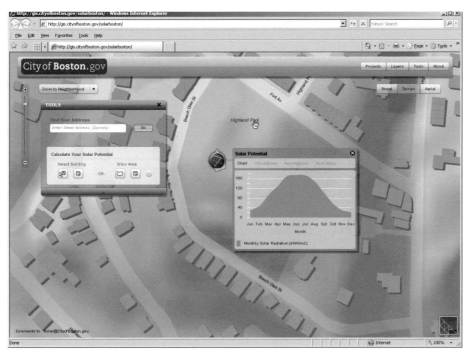

**Figure 2.11**    With the Solar Boston Web GIS application, users can select a rooftop or draw a polygon on the map, and then have the monthly solar energy potential calculated. The calculation is done by the server because of its complexity and the large volume of data needed.

Courtesy of Boston Redevelopment Authority.

## 2.4 USER EXPERIENCE DESIGN

The user experience has become an important consideration for Web application design, including Web GIS design. **User experience (UE or UX) refers to the level of satisfaction a person gets from using a product or a service.** Traditionally, GIS is most often used by professionals in the workplace, who have years of GIS training or experience and who know how to use GIS. Thus, professional GIS design has mostly focused on functionality and neglected improving the user experience. By unlocking GIS technology, the Web has delivered GIS to a large number of users who have no concept of GIS except that they expect it to be as easy to use as a regular Web page and as fun to use as a multimedia center. To meet user expectations, the Web should wrap necessary functionality with a pleasant user experience conveying the three main principles of UE design—making Web applications fast, easy, and fun to use. The first principle, speed, is mostly determined by the server and is closely related to Web service optimization (see section 3.5). The latter two principles are introduced in the following sections.

## 2.4.1 EASE OF USE

The most important factor in making a Web application easy to use is the maxim "Don't make me think!" (Krug 2000). This is in contrast to other mistaken criteria such as "no more than three clicks to content." What really counts instead of the number of clicks is how hard it is to click each time—that is, how long it takes to find the right place to click. Each click should be painless, unambiguous, and give you continued confidence that you are on the right track. Otherwise, a Web application can have a big learning curve that can turn users off.

Brian Noyle (2009), Clint Brown et al. (2008), and many others have outlined several points about how to make a Web GIS easy to use, including the following:

- **Hide complexity:** Web GIS should create intuitive and focused applications that serve a particular purpose, that provide as few tools and few layers as needed, and that don't flood users with unnecessary functions, features, and data layers that won't be used and that distract you by making you have to think about which tool to use. The terms should be self-explanatory, not GIS jargon.

- **Provide feedback:** Many GIS functions take multiple steps to complete. A good application should lead users through a well-defined workflow with visual cues, provide consistent and meaningful feedback in the user interface, and assure users that they are on the right track.

## 2.4.2 MAKING IT FUN

In the early 1990s, the world was impressed to see even a picture image appear on the Web. That kept users excited for a few years until they began to expect something even more impressive—with things like animation effects, compelling multimedia, and an intuitive interface. To keep pace with users' rising expectations, **rich Internet application (RIA)** technologies came into play. The term RIA was coined by Jeremy Allaire (2002) to describe the benefits of a new version of the Adobe Flash Player. Later, **RIA came to generally refer to Web applications that provide a rich and engaging user experience comparable to desktop applications.** RIAs are typically developed with AJAX, Flex, or Silverlight.

Higher user expectations and advances in RIA technology have resulted in many RIA-style Web GIS applications that use browser APIs such as ArcGIS APIs for JavaScript, Flex, and Silverlight. These applications can fetch map tiles for the current map extent and nearby areas and use asynchronous communications behind the scenes to provide users with smooth map pan and zoom. They can display operational layers vividly by using animation icons to simulate flying helicopters or flag terrorist attacks. Animation icons show situational details when the pointer hovers over a feature on the map. They can use pie or column charts to present demographic statistics for areas surrounding a new retail store. They can link locations on the map with photos of highway traffic and real-time videos of critical infrastructure. They can overlay maps on a video to simulate the path of an earthquake. They can play multiple temporal aerial photos in sequence to produce a movie effect showing historic land-use change. They can employ animation effects such as zoom, fade, flip, rotate, bounce, fly, ripple, sparkle, and many more—all demonstrating that GIS can be fun to use. These rich and engaging user experiences, when employed appropriately, can greatly increase user productivity and improve user satisfaction.

This chapter summarized the basic architectures, components, and programming technologies of Web applications and Web GIS applications. While not intended to be a "how to," the contents provide guidelines on design questions such as which basemap to use, whether to use one available on the Web or create it in-house, what the operational layers should be, how users should interact with them, what tools are needed, whether the processes should be done by the client or the server, and how to achieve a good user experience. Properly following these guidelines can lead to Web GIS applications that ensure good usability and fast performance.

## Study questions

1. What are the three foundational technologies of WWW?

2. What is the three-tier architecture of a Web application?

3. List the main Web server technologies and the main browser technologies.

4. What are the typical data formats for exchanging data between the Web server and Web client? Compare them.

5. What are the components of a basic Web GIS application?

6. What is a geobrowser? What is an online virtual globe? Give some examples.

7. What is a thin client and what is a thick client in the context of Web GIS?

8. What is the best strategy to partition workloads between a server and a client in a Web GIS application? Give a good example of such a Web GIS application.

9. What is user experience, and what are the main principles of user experience design?

10. What is RIA? Describe a Web GIS application that you think is RIA. What technology does it use to achieve RIA?

## References

Allaire, Jeremy. 2002. Macromedia Flash MX—A next-generation rich client. Macromedia white paper. http://www.adobe.com/devnet/flash/whitepapers/richclient.pdf (accessed August 20, 2009).

Brown, Clint. 2009. Bringing your geographic information to life. ESRI International User Conference, San Diego, Calif., July 13–17.

Brown, Clint, Jeremy Bartley, Ismael Chivite, Bernie Szukalski, and Rob Shanks. 2008. ArcGIS in a Web 2.0 World. ESRI International User Conference, San Diego, Calif., August 4–8.

Butler, Declan. 2006. Virtual globe: The Web-wide world. *Nature* 439: 776–78.

Foresman, Timothy W. 2008. Evolution and implementation of the digital Earth vision, technology, and society. *International Journal of Digital Earth* 1 (1): 4–16.

Fu, Pinde, and Paul M. Rich. 2002. A geometric solar radiation model with applications in agriculture and forestry. *Computers and Electronics in Agriculture* 37 (1): 25–35.

Gong, Jianya. 1999. *Contemporary GIS theory and technology.* Wuhan, China: Wuhan University of Surveying and Mapping Science and Technology Press.

Gosling, James. 1996. Java is fusion of several kinds of programming. Fifth International World Wide Web Conference, Paris, May 6–10. http://iw3c2.cs.ust.hk/WWW5/www5conf.inria.fr-webcast/gosling.htm (accessed November 16, 2009).

Knight, Greg. 2009. Boston showcases solar power potential with ESRI's Web GIS. ESRI speaker series podcasts. http://www.esri.com/news/podcasts/audio/speaker/user_knight.mp3 (accessed August 22, 2009).

Krug, Steve. 2000. *Don't make me think! A common-sense approach to Web usability.* Berkeley, Calif.: Pearson Technology Group.

Net Applications. 2009. Browser market share. http://marketshare.hitslink.com/browser-market-share.aspx?qprid=0 (accessed September 17, 2009).

Netcraft. 2009. August Web server survey. http://news.netcraft.com/archives/web_server_survey.html (accessed September 17, 2009).

Noyle, Brian. 2009. Usability and the GeoWeb, Parts 1, 2, and 3. GIS and .NET development blog. http://briannoyle.wordpress.com (accessed November 9, 2009).

Quinn, Sterling. 2008. Design patterns for Web maps. *ArcGIS Server Blog,* September 22. http://blogs.esri.com/Dev/blogs/arcgisserver/archive/2008/08/05/Design-patterns-for-Web-maps.aspx (accessed August 21, 2009).

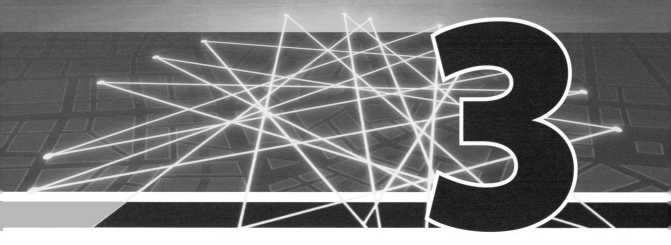

# GEOSPATIAL WEB SERVICES

In contrast with Web sites, which are usually isolated and closed to external software systems, Web services are considered open in that they are programming interfaces that can be accessed by other applications over the Web. Geospatial Web services leverage the power of GIS, programming components, and the World Wide Web to serve an expanding variety of needs. This synergy of technologies allows geospatial Web services to significantly change the way GIS is developed and used.

Web service technology has had a dramatic effect on the evolution of Web GIS products, leading to niche products such as server products, geobrowsers, and Web portals that correspond to the service, client, and broker roles in Web service architecture. A greater number of organizations are publishing data and functions in Web service format. These services are increasingly used as the building blocks in service-oriented architecture (SOA) or Web-oriented architecture (WOA, see chapter 7), which allows GIS to be seamlessly integrated in mainstream information systems. The use of Web services provides a new approach to interoperability. It bypasses the complexity of local data conversion and software installation and enables different systems to work together at the Web service level. Web service architecture naturally fosters collaboration among organizations and can act as the supporting framework for National Spatial Data Infrastructure (NSDI).

Web service technology is of profound importance to GIS today and into the future. This chapter introduces the basics of Web services. Web service applications are covered in many chapters of this book, especially chapter 4, in which Web services are described as the main type of resource for geospatial mashups, and chapter 7, which presents Web services in the context of SOA, WOA, and NSDI 2.0.

The basic knowledge covered in this chapter includes the concept, impacts, functions, and benefits of Web services. SOAP-based and RESTful Web services are introduced, as well as Web-service-related standards such as WMS, WFS, WCS, CSW, GML, KML, and GeoRSS. Web service optimization techniques for better performance, scalability, reliability, and security are also discussed.

## 3.1 FROM WEB SITES TO WEB SERVICES

The Palo Alto Research Center (PARC), then Xerox PARC, launched the first Web mapping application in 1993. This represented a significant milestone in the evolution of Web GIS. The first generation of Web GIS consisted mostly of stand-alone Web sites, where spatial visualization and query functions were available only to their own client browsers and couldn't be reused and integrated in other information systems. How to make them open to other applications over the Web became the operative question. The answer was the Web service technology that emerged in the late 1990s.

### 3.1.1 THE NEED FOR WEB SERVICES

An explosion of Web-based mapping applications followed the birth of Web GIS in 1993, and a bevy of Web GIS software products appeared on the market. These early Web GIS technologies had limitations, however, in both their internal architecture and in their integration with other information systems. Because of these limitations, Web GIS was underused, and its potential was not fully realized (Huang 2002):

- **Isolated systems, difficult to reuse and integrate:** Early Web GIS technologies were developed separately and seen as an "independent solution." They were isolated from other systems, and it was difficult to share information and functions between different systems or to incorporate Web GIS seamlessly into other IT systems. Systems that do not expose programming interfaces cannot communicate with each other (figure 3.1). In cases where functions and information in

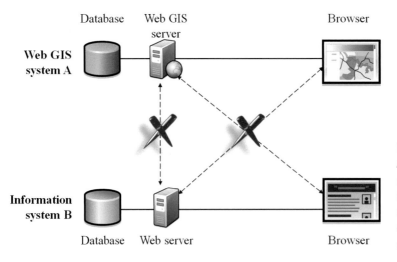

**Figure 3.1** Neither system A nor system B exposes a programming interface to the Web, so the client and server of one system cannot communicate with the other system.

one system are needed by another, information cannot be shared when neither system can call the other.

An example is an enterprise information system, such as an ERP (enterprise resource planning) system or a CRM (customer relation management) system, which needs maps to show where its resources or customers are, but the enterprise information system wouldn't be able to seamlessly embed the mapping and querying functions of a geospatial Web site if the site had no interface for the other information system to call.

- **Tightly coupled client and server, inflexible to change:** When changes or improvements need to be made to such Web GIS applications, the cost to break the tightly coupled modules and reengineer them is high. It is difficult for such systems to provide functioning components that can stand apart and be quickly remixed to satisfy the challenge of new demands.

GIS is an integral part of the information infrastructure of organizations, especially in the Internet era. The value of geospatial information is increasingly being recognized, and it is being used in the decision-making process in an increasing number of disciplines. GIS is a vital component of the mainstream information industry. Demand is strong for Web-based integration and reusable components, both to facilitate the sharing of information and integration of functionality and to achieve more flexible system architecture. The need for Web-based integration is prevalent not only in the geospatial sector, but in the IT industry as a whole. It has been an important research frontier for software vendors, institutes, and government agencies. By the late 1990s, organizations such as Sun Microsystems, Microsoft, Oracle, IBM, and the World Wide Web Consortium (W3C) had all declared their research in the subject, resulting in a new generation of Web technology—Web services.

The definition of Web services has evolved over the years. Early definitions of Web services were mostly tied to SOAP, XML, and WSDL (see section 3.3.1). With the considerable changes that have been made in Web service technology, SOAP is not the only way to implement Web services anymore. REST-style Web services (see section 3.3.2) have expanded the Web service concept to a more inclusive definition:

**A Web service is a program that runs on a Web server and exposes programming interfaces to other programs on the Web.**

Web service technology represents an important evolution of distributed computing. It replaces a local function with a function that is provided remotely by a Web server. Desktop software consists of a collection of smaller programs that operate together. Imagine these smaller programs distributed across the Web and running on different servers, but they still communicate with each other and function together as a whole—that's the idea of Web services. A comparison of Web pages and Web services can also help explain Web services:

- Web pages are for users to read and understand. They are in HTML format with a mix of contents and presentation styles (such as font, size, color, and position).

- Web services are for other computer programs to call. They are composed of Web-based programmable components, and their results are usually in XML, JSON, or other structured formats that computer programs can parse and reuse.

There are three roles that complete Web service architecture: the provider that provides the service, the consumer that uses the service, and the registry that acts as a broker between the provider and the consumer to facilitate the collaboration. The provider and the consumer don't necessarily know of each other's existence. The provider can register its availability and interface in the registry; and the consumer can search the registry, discover the service it needs, and then use the service (figure 3.2).

Web services are programming interfaces and Web programs, thus inheriting the characteristics of both. As such, they possess a number of advantages over more traditional approaches to computing:

- **Open to other software systems over the Web:** Web services can exchange information and share functions with other software systems over the Web, breaking the isolation status of early Web applications. This means developers can get the functions they need over the Web without having to install or rely on software local to their computers.

- **Independent of programming language and operating system:** A Web service is like a self-contained black box with only its programming interface exposed. Whatever programming language is used to implement a Web service (.NET, Java, or C++, etc.), whatever operating system it runs on (Windows or UNIX, etc.), and whatever Web application server it is deployed on (Apache, IIS, etc.), none of this affects how clients can consume the service. Developers have the freedom to choose whatever tools or programming language they like. Thus, Web services facilitate the ability to build composite applications based on heterogeneous services operating across many different platforms.

- **Loosely coupled with remixing on demand:** Web services and their clients are not tightly bound to one another. A Web service can be consumed by multiple clients, and a client can consume multiple Web services. A Web service and its clients do not need to run on the same server, and they do not need to be compiled together. If a Web service does not live up to its client's satisfaction, the client can easily switch to a different Web service with the same interface standard. A Web service can make any internal changes (e.g., migrating from Java to .NET, or the reverse)

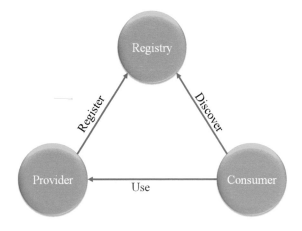

**Figure 3.2** This schematic shows the three roles in a complete Web service architecture and how they relate to each other.

without disrupting the client as long as the programming interface it exposes remains the same. The advantage of being loosely coupled is the flexibility to remix and deploy to meet new business needs.

- **One-for-all release and update:** When a Web service is updated or a new version is released, the change only needs to be made on the server side. As soon as the change is made, all clients consume the latest version. There is no need to run installation or an update on each client computer. This is a significant advantage of Web services over desktop programming components.

## 3.1.2 GEOSPATIAL INDUSTRY SUPPORT OF WEB SERVICES

Web services are the foundation of some important trends in information technology—service-oriented architecture and mashups. The geospatial industry is spearheading strategies to facilitate the authorship, publication, discovery, and consumption of GIS Web services through updates of Web GIS products, and this is clearly notable is the ESRI ArcGIS product family.

Corresponding to the provider, consumer, and registry roles in Web service architecture, Web GIS products have evolved into three main categories (figure 3.3):

1. **Server products** such as ArcGIS Server, which can publish geospatial Web services; serve maps, data, and analytic models for use across the Web; and enable the Software as a Service (SaaS) business pattern, in which companies in the geospatial industry provide software and services using Web services or Web applications.

2. **Geospatial Web portals (Geoportals)** such as ArcGIS Services Directory, ArcGIS.com catalog, and ESRI ArcGIS Server Geoportal Extension, which serve as brokers between the provider and consumer.

   - ArcGIS Server Web service metadata is automatically registered in ArcGIS Services Directory. The service metadata can be published in ArcGIS Online, ArcGIS Server Geoportal Extension, and many other geoportals (see chapter 6).

   - The metadata can also be indexed by general-purpose Web portals such as Google, Microsoft Bing Search, and Yahoo, which makes the services searchable by a bigger audience.

   - Users can go to the aforementioned portals to browse or search for geospatial Web services they need, and then connect and use them.

3. **Client products,** which can consume geospatial Web services with and without programming.

   - **Geobrowsers** such as ArcGIS Explorer, Microsoft Bing Maps, Google Earth, NASA World Wind, OpenLayers viewer, and Gaia allow users to consume a variety of geospatial Web services without having to write custom code (see section 2.2.3).

   - **Custom applications development** allows users to build custom Web, desktop, and mobile applications by integrating a series of Web services to meet their business needs. SOA is increasingly using Web services as the implementing technology. The recent addition of browser APIs, such as ArcGIS APIs for JavaScript, Flex, and Silverlight, greatly reduces the development effort and allows users to mash up Web services to create both casual and enterprise applications.

In addition to the new products that can serve, catalog, and consume geospatial Web services, desktop software such as ArcGIS Desktop has evolved as well. While still holding an important role in the standard workflow for desktop GIS users, ArcGIS Desktop takes a new lead in the Web GIS-centric era as a tool to prepare GIS resources for Web service publishing. In addition, it can consume many types of Web services. With its ability to combine remotely hosted services with locally available data and capabilities, ArcGIS Desktop becomes a powerful tool that supports the Software plus Services (S+S) architecture (see section 7.4.2).

### 3.1.3 THE BENEFITS OF GEOSPATIAL WEB SERVICES

Web services have become the heart of GIS, representing significant progress in distributed GIS. They provide a means to connect a range of heterogeneous applications and connect a network of distributed computing nodes, ranging from servers and desktop clients to mobile clients. The technology, which has exercised a positive influence on the geospatial industry, will help to realize

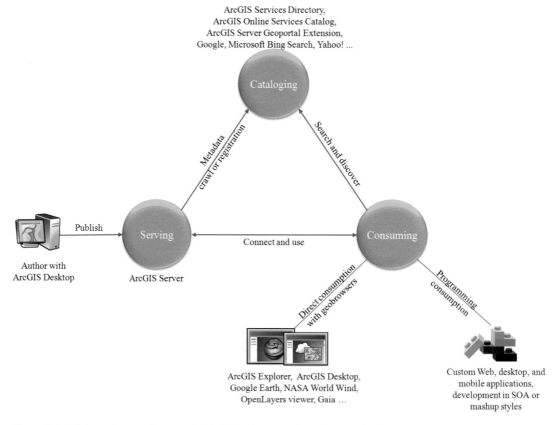

**Figure 3.3** Web services are the core of Web GIS. Web GIS products have evolved to support geospatial Web services in authoring, publishing, cataloging, and consuming services.

many of the shared visions of geospatial information sharing, interoperability, and collaboration that have been formulated over the last decade.

## FOSTERING MANY NEW NICHES IN THE GEOSPATIAL INDUSTRY

Geospatial Web services provide many new business niches, whether you're on the server side, client side, or cataloging side.

**On the server side:** If you own a large amount of data, you can be a data and mapping service provider. If you have certain models and expertise, you can publish them as specialized geoprocessing services. The services can be offered for free or at a per-use price. The range and diversity of Web services will bring on a wealth of choices for Web GIS users.

**On the client side:** If you don't have GIS data and tools, you may choose to develop client products such as geobrowsers and focus on usability, performance, and the various service types they support. You can work in consulting and development, combining available Web services to meet the needs of your customers, even if they don't have local data or GIS software.

**On the catalog side:** You can be in the geospatial Web portal business (see chapter 6), collecting the information provided by Web services for a certain region, a certain theme, or within a certain standard; cataloging the information; and making it discoverable for those who need it.

## BUILDING BLOCKS FOR USE IN MAINSTREAM INFORMATION SYSTEMS

It's widely acknowledged that geospatial technology can greatly benefit nongeospatial industries once GIS is easily and seamlessly integrated into other information systems, but that potential has not yet been fully reached. The complexity of having to copy GIS data locally and to install GIS software components locally remains a barrier in many situations to using GIS.

Geospatial Web services, on the other hand, hide the complexity of GIS data and functionality, leaving it to be handled remotely on other servers, while exposing a Web programming interface for easy integration. This means that IT can simply access mapping, data, and geoprocessing Web services from a variety of sources without having to deal locally with the geospatial complexity. Web services allow GIS and spatial processing to be integrated flexibly and extensively with other IT business systems, such as ERP and CRM, serving as the best building blocks for SOA. This openness and flexibility can greatly expand the GIS market.

## A NEW APPROACH TO ACHIEVING INTEROPERABILITY

The ability of diverse systems to work together is one of the most important challenges facing GIS. Before Web services technology was available, interoperability was mostly accomplished at the data formatting level. Toward that end, standards bodies define open data formats that are vendor neutral. Different vendor software needs to be able to either read or write data formats directly or import and export such formats. This procedure usually requires users to make duplicate offline data copies and install tools locally. It makes it difficult to achieve real-time flexible interoperability over the Web platform.

With the emergence of Web services, data and functions can be exposed as Web services and linked "on demand" to build solutions. This capability gives the GIS industry the ability to move beyond data conversion and converter installation into Web-service-based interoperability (Bacharach 2005). Realizing this opportunity, standards bodies such as the Open Geospatial Consortium (OGC) and International Organization for Standardization (ISO) have defined a series of Web service standards. With these standards, GIS applications are not tied to a specific software vendor. Organizations can manage data using the methods and formats best suited to their needs while exposing Web service interfaces that conform to specific open standards. Other users can use these services regardless of which vendors are behind the services (see section 3.3).

### AN IMPORTANT FRAMEWORK FOR REALIZING SPATIAL DATA INFRASTRUCTURE

Spatial data infrastructure (SDI) refers to the technology, policies, standards, human resources, and activities necessary to acquire, process, distribute, use, maintain, and preserve spatial data. The keys to implement regional, national, and global SDIs are standards, sharing, collaboration, and coordination. Web service architecture establishes a specific type of relationship between service providers and information consumers that supports the dynamic integration of data and functions, which is the key to creating an SDI.

With Web services, data sharing and collaboration no longer means having to make do with an offline copy or having to download online. Members of the GIS community can use distributed data management and integrate versatile GIS services. Geospatial data can be left in the hands of a responsible custodian, where it can be maintained to keep its integrity and currency. The data can then be shared as Web services or as a component of processing services for other organizations to use. A good example is a local government that can continuously maintain and update its land records while serving them to other organizations via Web services. Rather than maintaining its own copy of basemap data, a utility company can directly use a local government's basemap and serve its facilities data back to the government for use in permitting and land-use planning. This type of synergy will serve as the backbone for a whole new era of collaboration among organizations (Dangermond 2008).

## 3.2 GEOSPATIAL WEB SERVICE FUNCTIONS

Geospatial Web services can be categorized by the functions they provide: map services, data services, analytical services, and metadata catalog services. This section explains what each type of service can provide in terms of functionality. ArcGIS Server services are used as examples when applicable.

### 3.2.1 MAP SERVICES

Map services allow clients to request maps for a specific geographic extent, and the maps are returned in an image format (e.g., JPEG, PNG, or GIF). This is the most common type of geospatial Web service. Beyond mapping and map viewing, map services may also support the attribute query, spatial identify, and dynamic reprojection functions. Map services can either be cached or dynamic. A map service that fulfills requests with precreated tiles from a cache is called a cached or tiled map service (see section 3.5.1). A cached map service can significantly improve performance time in delivering maps and is typically used to serve basemaps or maps where the content is relatively static and changes very little over time. A dynamic map service requires the server to render the map each time a request comes in. Dynamic map services are typically used to serve maps whose data is constantly changing or to serve maps with operational or theme layers.

Map services can be 2D or 3D. Throughout the book, you will see many figures representing examples of 2D and 3D map services. A 3D map service, also called a globe service, can improve communication by providing a more realistic spatial view. This virtual globe allows users to zoom and spin it for an enhanced user experience. Well-known 3D map services include those of Google Earth, Microsoft Bing Maps 3D, and ArcGIS Online globe services. Server products that can publish a globe service give users the ability to publish their own basemap in 3D with up-to-date, focused content that meets specific needs. 3D services can provide a portrait of the natural terrain using elevation as the third dimension (e.g., figures 1.8 and 10.1); illustrate the urban skyline using building height as the third dimension or by using a more realistic 3D model (figure 3.4); or emphasize a certain theme using a particular attribute value as the third dimension (figure 3.5).

## 3.2.2 DATA SERVICES

Data services allow you to query, edit, and synchronize data over the Web. Some data services are also map services, which let you see the map display as well as have access to raw data.

### FEATURE EDITING SERVICES

A feature editing service enables Web editing, which allows end users to add, edit, and delete features in the geodatabase remotely via a Web GIS client application. End users can use a set of simple sketch tools to draw new features or reshape existing features directly on a Web map. This includes splitting or merging existing polygons, or trimming or extending existing lines. Users can also edit attributes (figure 3.6) and add attachments such as digital photos.

Feature editing services, Web editing, and geographic sketching facilitate public participation GIS (PPGIS) and volunteered geographic information (VGI) (see section 10.1). These capabilities allow organizations to extend spatial data editing to a larger, more diverse audience, provide venues for online communities to become active contributors to the geodatabase, and can be used to capture user-generated content (UGC). Feature editing services can support efficient GeoDesign (see chapter 1) by allowing users to quickly sketch their design ideas on top of a digital map and share their ideas with the community or other team members. These sketches can serve as the basis for collaborative design and be used for suitability or impact analysis.

### SEARCH SERVICES

A search service is for indexing and searching GIS resources (e.g., a data layer, a table, or a whole enterprise geodatabase). Web users, especially users within an organization's intranet, can search for desired GIS resources by querying the search service, for example, by using keywords.

**Figure 3.4** A globe service can portray building structures on top of ground terrain such as this 3D globe service of Philadelphia's urban landscape.

Courtesy of Pictometry International Corp. and U.S. Geological Survey.

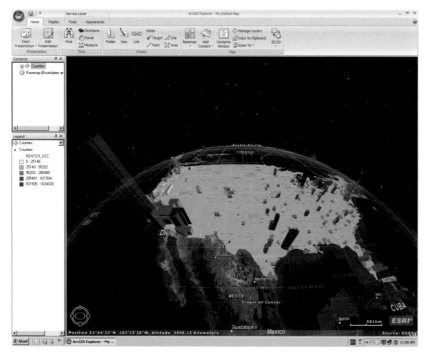

**Figure 3.5** A 3D map service can emphasize a certain theme using an attribute as the third dimension. This 3D map service uses renter-occupied household units per county as a unit of elevation to show how housing rentals are distributed.

Courtesy of U.S. Geological Survey and U.S. Census Bureau.

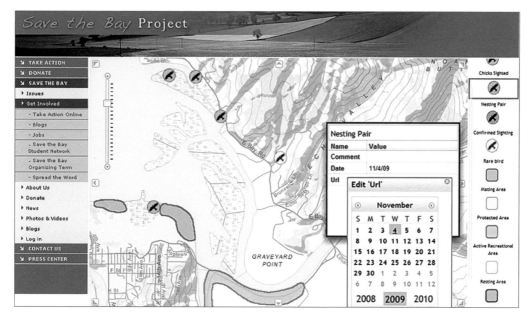

**Figure 3.6** Feature editing services enable Web editing, which allows end users to add, edit, and delete spatial features and their attributes. For example, citizens can mark a preservation map indicating when and where they see an endangered bird and note what kind it is.

Courtesy of U.S. Geological Survey.

A search service is different from a metadata catalog service (introduced later in this section). While both are for discovering GIS resources, a search service indexes the data, and especially the attribute tables, directly, while a metadata catalog service indexes the metadata.

## IMAGE SERVICES

An image service provides access to raster data, such as remote sensing imagery and digital elevation data, through a Web service. It supports raw data extraction and download, and often map display as well. For example, ArcGIS Server Image Extension allows organizations to directly publish large image collections without preprocessing as image services. Image services can perform dynamic mosaicking and on-the-fly processing and create multiple image products from a single source. This makes it easier and quicker to distribute very large volumes of rasters to a range of client applications.

## GEODATA SYNCHRONIZATION SERVICES

A geodata synchronization service is used to periodically replicate or synchronize data updates between different geodatabases over the Internet. This type of service is particularly useful in situations of distributed geodatabases in different locations—at different levels of government, for instance.

### 3.2.3 ANALYTICAL SERVICES

Analytical geospatial Web services perform a variety of GIS analysis functions, from commonly used geocoding and road network analysis to geometry transformation and geoprocessing services. Geoprocessing services can accommodate any workflow and execute any tasks that you design.

#### GEOCODING SERVICES

Geocoding is the process of converting street addresses to geographic coordinates, usually latitude and longitude. This is not always a straightforward process and can be complicated by ambiguous addresses or addresses that do not exist in the database. Thus, geocoding results are usually a list of possible matches. A geocoding service, also referred to as a locator service, exposes the geocoding function to the Web. In addition to geocoding, the service may also support reverse geocoding, which is the process of finding the address pertaining to a geographic location.

A geocoding service supports a wide range of applications, from finding a customer's shipping location to finding a stop in routing analysis. Geocoding allows you to find an address on a map and see how it relates to surrounding features. There are many available geocoding services, such as ones from Microsoft, Google, Yahoo, ArcGIS Online, and MapQuest. In case their address information for your area is not up-to-date, your address formatting is different, or you want people to find addresses by their local names and aliases, you can use products such as ArcGIS Server to create your own geocoding service.

#### NETWORK ANALYSIS SERVICES

Here, "network" refers to transportation network systems such as streets and highways. Network analysis services can support the following functions:

- **Routing:** Gives directions for the shortest or fastest route between two or more stops. Routing services can account for speed limit, U-turn policy, and impedance. The impedance factor attempts to account for traffic congestion, traffic lights, speed bumps, and other traffic delays.

- **Calculating service area:** Calculates a network service area, which is a region that encompasses all accessible streets within a specified driving time. For instance, the five-minute service area for a point includes all the streets that can be reached within five minutes from that point location. Service areas can help users evaluate site accessibility—for example, they can help planners to choose the best site for a fire station so that it can cover the optimal area within a few minutes, or help managers to select the best spot for a new store so that it can serve the most customers in a region (see figure 4.9C).

- **Finding the closest facility:** Finds the closest facilities to a location based on driving time or distance. This is especially useful for location-based services such as finding the restaurants or post offices near a mobile phone user.

#### GEOMETRY SERVICES

A geometry service can perform geometric transformation and calculations such as buffering, simplifying, merging, splitting, calculating areas and lengths, and projecting coordinates.

## GEOPROCESSING SERVICES

A geoprocessing service provides the flexibility to share the functions or models you author locally—for example, in the ArcGIS Desktop geoprocessing framework—with the Web. The functions can range from local community planning to global climate change researching, and from simulating the past to predicting the future. Examples in this book include solar potential calculation (see figure 2.11), the plume model for a chemical spill (see figures 1.8 and 5.14), viewshed calculation (see figure 4.9A), and terrain profile (see figure 4.9D). The Ecosystem Service Modeling System (EcoServ) by USGS, USDA, NASA, the Chinese Academy of Sciences, and a number of universities aim to build an integrated spatially explicit ecosystem service model to simultaneously monitor, quantify, and predict the changes of ecosystem goods and services at landscaping to national levels. The current research focuses on models related to water resources, wildlife, plants, soils, and greenhouse gas (GHG) emissions. The modeling system will be shared on the Web as geoprocessing services to accommodate a variety of users and purposes (EcoServ Wiki 2010).

## 3.2.4 METADATA CATALOG SERVICES

Metadata (data about data) can describe GIS data and services. A metadata catalog service allows publishing and searching metadata. It facilitates sharing geospatial information and services. For example, a provider can publish metadata about their data and services for others to discover, while a user can query the metadata service to discover what data and services fit particular needs. Products that support catalog services include ArcGIS Server Geoportal Extension (see chapter 6).

# 3.3 WEB SERVICE TYPES

The communication between a Web service and a client involves the client sending requests to the Web service, and the Web service returning responses to the client. Depending on the format of communication used, Web services can be categorized into two main types—namely, SOAP-based Web services and REST-style Web services. This section explains what these services are, how they work, how they differ, and industry trends. Examples of geospatial Web services are included.

It should be noted that while this section focuses on SOAP and REST, Web services are not confined to these two types. All Web programs that communicate with their clients through structured data formats, such as POX (Plain Old XML, as opposed to SOAP-encapsulated XML), CSV (comma-separated value), binary maps, or other formats that can be automatically parsed by computer programs, should also be considered Web services.

## 3.3.1 SOAP-BASED WEB SERVICES

SOAP, which originally stood for Simple Object Access Protocol, is a protocol specification for exchanging structured information in XML format. SOAP became a W3C recommendation in 2003, but "Simple Object Access Protocol" was considered misleading, so it was dropped by W3C in 2007. **SOAP-based Web services use HTTP Post to send requests. The requests and responses are in SOAP-encapsulated XMLs** (figure 3.7).

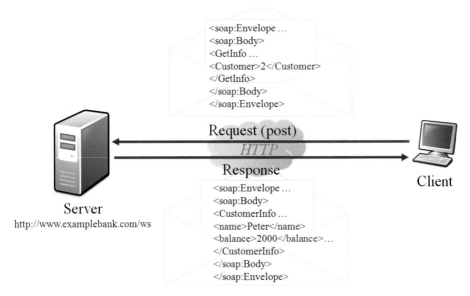

**Figure 3.7** SOAP-based Web services rely on HTTP Post and SOAP-encapsulated XMLs to pass requests and responses between the client and the server.

A SOAP message packages an XML body in an XML envelope. This "XML in XML" structure makes it difficult to construct and parse manually. Fortunately, there are tools that do this for you. SOAP is often used in conjunction with WSDL (Web Service Description Language). WSDL is an XML-based language for describing Web services and how to access them. Developer environments, such as those for .NET and Java, provide SOAP toolkits that can automatically generate sets of local APIs. The local APIs will construct and parse the SOAP XMLs for you.

For example, ArcGIS Server has been providing SOAP-based Web Services since 2004. Some clients, including ArcMap, ArcGIS Explorer Desktop, and applications developed with ArcGIS Server Web ADF communicate with ArcGIS Server via the SOAP interface. Users can also build customized GIS applications by programming against the SOAP WSDL. Figure 3.8 shows the WSDL of an ArcGIS Server map service and the programming environment generating local APIs using the WSDL.

### 3.3.2 REST-STYLE WEB SERVICES

REST (Representational State Transfer) is a style of software architecture introduced in 2000 in the doctoral dissertation of Roy Fielding (2000), one of the principal authors of the HTTP specification. This architecture style is designed to fully take advantage of HTTP, while reducing system complexity and improving system scalability (Richardson and Ruby 2007).

In the strictest sense, REST refers to a collection of network architecture principles that outline how resources are defined and addressed. The principles of REST are abstract and there are variations in their interpretation. These principles are not fully observed by many of the Web services

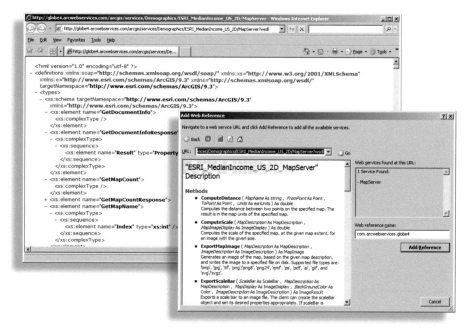

**Figure 3.8** Web services published by ArcGIS Server have both SOAP (illustrated here) and REST interfaces. On the upper left is part of a map service WSDL. At the lower right, local APIs are generated in Microsoft Visual Studio.

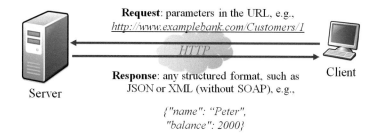

**Figure 3.9** In the most common implementation of REST, all requests are in the form of a URL. The request URL has all the necessary parameters, and the response can be in JSON or some other format.

that claim to use REST style. Based on the reality of how REST is implemented in the IT industry, REST-style Web services can be defined as the following:

**RESTful Web services are Web services that transmit data over HTTP without an additional messaging layer such as SOAP. In the most common RESTful Web service architecture, the client sends all parameters in the request URL.**

In the most common implementation of REST, all requests are made by a URL, and all parameters are in the URL. REST does not define standards for the server response format, but JSON and XML (without SOAP encapsulation) are frequently used. Figure 3.9 illustrates the same request and response in REST style as in figure 3.8, which uses SOAP.

RESTful Web services are characterized by a number of underlying principles. The main ones are as follows:

## 1. Identification of resources through URIs (or URLs)

Everything is a resource, and every resource gets a URI (Uniform Resource Identifier, or URL for easier understanding). The resources and their URLs are organized in a hierarchy that is intuitive, predictable, and easy to understand. Without much documentation, developers can easily construct a URL to locate and access the resources they need.

---

The ArcGIS Server Web services provide a hierarchy of resources. For example, the URL for all Web services available on an ArcGIS server is

```
http: //server.arcgisonline.com/ArcGIS/rest/services
```

The URL for a map service named "ESRI_StreetMap_World_2D" is the above URL, plus the service name and the service type (e.g., MapServer).

```
http: //server.arcgisonline.com/ArcGIS/rest/services/
ESRI_StreetMap_World_2D/MapServer
```

The URL to identify the first data layer in this map service is the above URL, plus "0" (the layer count starts from 0).

```
http: //server.arcgisonline.com/ArcGIS/rest/services/
ESRI_StreetMap_World_2D/MapServer/0
```

---

## 2. Manipulation of resources through their representations

A resource can be returned in many formats such as JSON, PNG, XML, or SVG (scalable vector graphics). These formats are representations of the resource. A resource can have multiple representations. RESTful Web services allow clients to indicate which representation of a resource they wish to receive.

ArcGIS map services can respond in several representations. Using the following ArcGIS Online map service as an example,

`http://sampleserver1.arcgisonline.com/ArcGIS/rest/services/Portland/ ESRI_LandBase_WebMercator/MapServer`

The response can be in the following formats according to the parameter "f" in the request URL:

- f=html: The response is an HTML page.
- f=json: The response is a JSON object. This format is used primarily by ArcGIS APIs for JavaScript, Flex, and Silverlight (see chapter 4).
- f=help: The response is a context-sensitive Help document that provides information about the resource.
- f=lyr: The response is a layer file for viewing in ArcMap.
- f=nmf: The response is a layer file for viewing in ArcGIS Explorer Desktop.
- f=jsapi: The response is a map viewer using the ArcGIS API for JavaScript. This map viewer allows you to dynamically interact with the map service.
- f=ve: The response is a Web page for viewing the map service interactively in a Web browser using Microsoft Bing Maps.
- f=gmaps: The response is a Web page for viewing the map service interactively in a Web browser using Google Maps.
- f=kmz: The response is a KML document wrapped in a KMZ file. It may be a footprint— that is, the spatial extent of a Web service—or the result of an operation.

REST also advocates the uses of a generic interface, especially different HTTP methods, to support the CRUD (create, read, update, and delete) operations:

- HTTP **Post**: create a resource
- HTTP **Get**: read a resource
- HTTP **Put**: update or create a resource
- HTTP **Delete**: delete a resource

### 3. Self-descriptive representations

Each client request and server response is a message. RESTful applications expect each message to be self-descriptive. This means that each message should contain all the information necessary to finish an operation. Each request and response cycle is "stateless" or "context-free" and is independent from the ones before and after it. Server applications do not need to check the context, so they do not carry the burden of a session. Stateless requests can be easily transferred from one server to another in a server farm deployment, which can greatly benefit load balancing and failover. This principle allows the use of server caches and browser caches, which tends to improve system performance.

**4. Hypermedia as the engine of application**

To accomplish a mission, the client needs to navigate from resource to resource and trigger the state transfers to complete the workflow. The server should provide link information that can help the client navigate through the resources and state changes.

The interpretation of these principles varies. In actuality, many Web services that claim to be "RESTful" have not fully complied with or implemented these principles. Most RESTful Web services simply use HTTP Get (i.e., put all parameters in the request URL), as indicated in the preceding definition. **"With REST, everything is a URL"** is the common slogan for REST. This is because HTTP Get is the simplest way to send requests over the Web. This appealing simplicity is what is making RESTful Web services so popular. REST is considered **"the command line of the Web."** You can put the URL request in a browser and see the results right in front of you.

Here are a few examples. The URLs here are not HTML encoded, to make your reading easier.

---

Using the ArcGIS Server REST API to generate a map covering the United States, asking the return to be a JPEG image of 800 pixels by 500 pixels, the request is the following:

`http://server.arcgisonline.com/ArcGIS/rest/services/`
`ESRI_ShadedRelief_World_2D/MapServer/export?`

> bbox=-185.33, 15.20, -9.53, 74.08
>
> &size=800, 500
>
> &format=jpg
>
> &dpi=96
>
> &f=image

Using the ArcGIS Server REST API to query a map service to obtain the median household income of every county in California, asking the return to be in JSON format, the request is the following:

`http://server.arcgisonline.com/ArcGIS/rest/services/Demographics/`
`ESRI_MedianIncome_US_2D/MapServer/0/query?`

> where=STATE_NAME='California'
>
> &outFields=MEDHINC_CY
>
> &f=pjson

Using Google's REST API to search for Web sites related to "GIS" (the default return format is JSON), the request is the following:

`http://ajax.googleapis.com/ajax/services/search/web?v=1.0&q=GIS`

Using Yahoo's REST API to geocode the address "701 First Ave, Sunnyvale, CA" (the default return format is XML), the request is the following:

`http://local.yahooapis.com/MapsService/V1/geocode?`

> appid=yourApplicationId
>
> &street=701+First+Ave
>
> &city=Sunnyvale
>
> &state=CA

### 3.3.3 COMPARING SOAP AND REST

In 2002, Amazon released its e-commerce Web services. Aware of the "REST versus SOAP" debate, Amazon provides both SOAP and REST interfaces to its Web services. By 2004, the platform had attracted more than 50,000 developers, with the preference being to use the REST method. In fact, 80 percent of the calls to Amazon's Web services were REST-based, with just 20 percent SOAP-based (Greenfield and Dornan 2004), an ongoing trend. **In many cases, the simplicity and efficiency of using REST outweighs the rigorous discipline of SOAP and the complexity in introducing SOAP-based Web services** (Fu et al. 2008). The advantages of using REST over SOAP generally include the following:

- **For producers:** lowering the cost of creating a service and lowering the overhead for hosting and supporting the service

- **For users:** lowering the learning curve and reducing the time and money needed to build GIS applications

- **For managers:** providing highly desirable architectural properties such as scalability, performance, reliability, and extensibility.

It is important to note that SOAP Web services are robust and comprehensive but complicated. REST Web services are simple and efficient but may not have all the capabilities of SOAP (Cappelaere et al. 2007). **Both SOAP and REST have advantages, and both have good prospects for future use** (table 3.1).

|  | SOAP | REST |
|---|---|---|
| **ADVANTAGES** | Rigorous and disciplined, mature, with many standards, better security support | Lightweight, simple, flexible, and efficient; can utilize HTTP cache; better support for browser clients such as JavaScript |
| **DISADVANTAGES** | Low efficiency (SOAP-packaged XML messages are heavy to transfer and large to parse), does not fully utilize HTTP advantages such as header and cache | Fewer standards, no formal way to describe a service interface |
| **MARKET** | Big IT infrastructure, large corporate | From small to big IT infrastructure, mass market |
| **PROGRAMMERS** | Professional developers | From nonprofessional to professional developers |

**Table 3.1**  A comparison of RESTful and SOAP-based Web services

Early on, it appeared that SOAP would become the definitive way to access a Web service. With the addition of REST-style Web services, developers and managers need to decide which method to use when they produce or consume Web services. For example, ArcGIS Server produces both SOAP and REST Web services (figure 3.10). SOAP is typically used for server-side development using .NET, Java, or ArcGIS Server Web ADF. ArcGIS Desktop and ArcGIS Explorer Desktop also communicate with ArcGIS Server via SOAP Web services. The REST interface is typically used for browser-side development. When you use ArcGIS APIs for JavaScript, Flex, and Silverlight, you are using the REST interface.

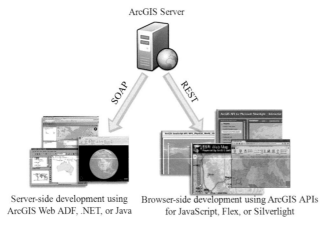

ArcGIS Server

SOAP        REST

Server-side development using
ArcGIS Web ADF, .NET, or Java

ArcMap and ArcGIS Explorer
Desktop

Browser-side development using ArcGIS APIs
for JavaScript, Flex, or Silverlight

ArcGIS Flex viewer, ArcGIS Explorer Online,
and other map viewers using browser-side APIs

**Figure 3.10** ArcGIS Server produces Web services in both SOAP and REST interfaces. ArcMap, ArcGIS Explorer Desktop, and server-side development typically use the SOAP interface, while browser-side development commonly uses the REST interface.

## 3.4 INTEROPERABILITY AND GEOSPATIAL WEB SERVICE STANDARDS

You may have heard the buzzwords "enterprise system," "digital city," "e-government," "federated architecture," and "spatial data infrastructure." These systems and business operations usually involve GIS data and tools that come from different vendors. Interoperability is required to make such data and software work together smoothly. The GIS industry has increasingly moved toward Web-service-based interoperability.

This section introduces Web-service-based interoperability and Web-service-related standards, including WMS, WFS, WCS, CSW, WPS, KML, GeoRSS, and GML. These standards are progressively being supported by the geospatial industry. For example, ArcGIS Server and GeoServer can serve a number of these standard services. On the client side, ArcGIS Explorer, ArcMap, Google Earth, NASA World Wind, Carbon Project Gaia, OpenLayers Viewer, 52° North OX–Framework Rich Client, and Skyline TerraExplorer can use many of these services.

### 3.4.1 WEB-SERVICE-BASED INTEROPERABILITY

**Interoperability** is the "capability to communicate, execute programs, or transfer data among various functional units in a manner that requires the user to have little or no knowledge of the unique characteristics of those units," or more specifically, is "software components operating reciprocally (working with each other) to overcome tedious batch conversion tasks, import-export obstacles, and distributed resource access barriers imposed by heterogeneous processing environments and heterogeneous data" (OGC 2007a).

Interoperability has been a main challenge for GIS. Over the years, the concepts, standards, and technology for implementing GIS interoperability have evolved through six stages (figure 3.11): (1) data converters, (2) standard interchange formats, (3) open file formats, (4) direct-read APIs,

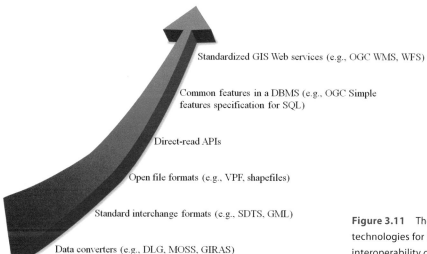

Standardized GIS Web services (e.g., OGC WMS, WFS)

Common features in a DBMS (e.g., OGC Simple features specification for SQL)

Direct-read APIs

Open file formats (e.g., VPF, shapefiles)

Standard interchange formats (e.g., SDTS, GML)

Data converters (e.g., DLG, MOSS, GIRAS)

**Figure 3.11** The evolution of technologies for implementing GIS interoperability covers six stages.

(5) common features in a database management system (DBMS), and (6) integration of standardized GIS Web services (ESRI 2003). Each of these strategies continues to play a significant role such that Web-service-based interoperability is increasingly being adopted in modern IT frameworks. The advantage is that organizations are liberated from tedious data conversion tasks and complex local tool installation. They can manage data and conduct analysis using whatever products they choose. As long as the Web service interfaces exposed conform to the agreed standards, other users can use these services without problem.

Standards are an important way to achieve interoperability. Standards specify the interface that different vendors should use. Vendors can develop products against the standards independently, without knowing each other, and their resultant products will still be interoperable. It's the same principle as power plugs and outlets, which conform to the same specification of shape and voltage, so that all you have to do is "plug and play."

**Web service standards** in essence specify the format of HTTP requests and HTTP responses, including what parameters should be in a request, the name of each parameter, what type of value(s) is expected for each parameter, and what type of results should be in the response. Web service standards facilitate interoperability and ensure the smooth flow of geospatial contents from different vendors over the Web (figure 3.12).

The main standards bodies related to geospatial Web services are as follows:

- **OGC:** Founded in 1994, OGC is a nonprofit international consensus-based standards organization leading the development of standards for geospatial services. OGC launched the OWS (OGC Web Services) Interoperability Initiative, which aims to establish a standard framework for GIS applications that can be seamlessly integrated in a variety of Web applications, location-based services, and mainstream IT. Its WMS, WFS, WCS, CSW, WPS, GeoRSS, and KML standards will be introduced later in this section.

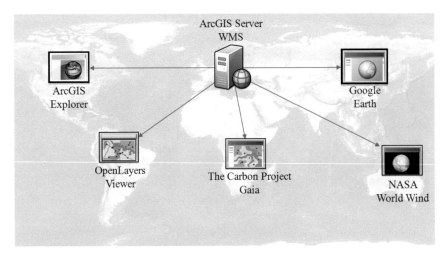

**Figure 3.12**  Interoperability relies on standards. Using OGC WMS standards as an example, a WMS published by ArcGIS Server can be consumed not only by ESRI products, but also by Google Earth, NASA World Wind, OpenLayers, and Gaia. Each client knows what parameters to send in the request and what response to expect.
Courtesy of U.S. National Park Service.

- **ISO/TC 211 (International Organization for Standardization, Technical Committee 211):** Also founded in 1994, ISO/TC 211 is a standard technical committee formed within ISO, which is responsible for covering the areas of digital geographic information and geomatics. The standards and specifications prepared by TC 211 are numbered, starting at 19101. Typically, ISO standards are more abstract while OGC standards are more specific. ISO/TC 211 works closely with OGC. OGC WMS and GML standards are ISO standards as well.

- **W3C:** The World Wide Web Consortium (W3C) is the main international standards organization for the Web. It was founded in 1994 by Tim Berners-Lee, the "father of WWW." W3C aims to address issues of incompatibility that may arise in Web technology by different vendors. W3C standards for HTML, XML, XHTML (Extensible Hypertext Markup Language), CSS, XML Schema, RDF (Resource Description Framework), SOAP, and WSDL are widely used and form the basis of geospatial Web services. The W3C SVG (scalable vector graphics) standard defines an XML format for two-dimensional vector graphics that is used to display maps in a variety of Web browsers and mobile devices. The W3C Basic GeoRSS standard is also used widely, though it is becoming obsolete.

### 3.4.2 WEB SERVICE STANDARDS

This section introduces the main Web service standards, including OGC standards for WMS, WFS, WCS, CSW, OpenLS, and WPS. You can visit the OGC official Web site (http://www.opengeospatial.org) to find detailed specifications as well as the list of OGC certified server and client products that comply with these standards.

### WMS

WMS (Web Map Service) is an OGC and ISO standard for requesting and serving maps over the Internet. Maps produced by WMS are generally rendered in a pictorial format such as PNG, GIF, or JPEG. WMS specifies the following operations (OGC 2006a):

- **GetCapabilities:** Requests service metadata from the server. The metadata, which is in XML format, describes the service title, area covered, operations supported, data layers and layer properties, coordinate system, abstract, keywords, contact information, and so forth.

> For example, you can retrieve information on the capabilities of the USGS National Map WMS by using the following GetCapabilities request:
>
> `http://nmcatalog.usgs.gov/catalogwms/base?request=getCapabilities`

- **GetMap:** Requests a map from the server. The request includes parameters such as the geographic layer(s), area of interest, output coordinate system, output map dimensions, and format. The request also can specify the styled layer descriptors (SLD) in which each layer is to be rendered. The response is usually an image stream.

> For example, you can use the following GetMap request to have the USGS National Map WMS generate a map in PNG format:
>
> `http://nmcatalog.usgs.gov/catalogwms/base?VERSION=1.1.1&REQUEST=GetMap&SRS=EPSG:4326&WIDTH=829&HEIGHT=326&BBOX=-180,17,-40,72&LAYERS=T_4,T_3&STYLES=,&FORMAT=image/pngh`

- **GetFeatureInfo:** Retrieves information for a location on a map. The request parameters are similar to the parameters used in the GetMap request and add the layers to be queried and the feature format to be returned. The response is usually in GML format and may include both geographical coordinates and attributes of selected layers in that location.

WMS is broadly used by many organizations. For example,

- USGS provides a series of WMS services, such as the National Map WMS (figure 3.13), National Atlas Transportation WMS, National Hydrography Dataset WMS, and U.S. Active Fire WMS.

- NASA Earth Observations (NEO, `http://neo.sci.gsfc.nasa.gov/Search.html`) offers an ongoing rich set of global earth science maps focused on climate and environmental changes via WMS (`http://neowms.sci.gsfc.nasa.gov/wms/wms`) and other means. Users can access time series satellite images on topics such as sea surface temperature, land surface temperature, vegetation index (the greenness of the Earth's surface), total rainfall, active fires, carbon monoxide concentrations, land-cover classifications, and incoming solar radiation.

**Figure 3.13**  A map viewer developed with the ArcGIS API for JavaScript is used as a client to display the National Map WMS published by USGS.

Courtesy of U.S. Geological Survey.

## WFS

WFS (Web Feature Service) specifies an OGC Web service standard for reading and writing geographic features in vector format. With WFS, clients can perform operations, including insert, update, delete, and query for geospatial feature data residing on the server. WFS defines the following main operations (OGC 2005):

- **GetCapabilities:** Requests service metadata. The response is an XML that describes service content and capabilities, including the feature types it can serve and the operations it supports.

- **DescribeFeatureType:** Requests the structure of the feature type that WFS supports.

- **GetFeature:** Retrieves a geographic feature and its attributes to match a filter query.

- **LockFeature:** Requests the server to lock on one or more features for the duration of a transaction.

- **Transaction:** Requests the server to create, update, and delete geographic features.

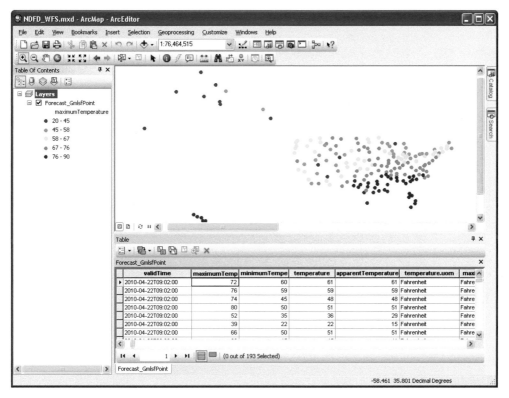

**Figure 3.14** ArcMap is used to display the National Digital Forecast Database WFS provided by the U.S. National Weather Service Meteorological Development Laboratory.

Courtesy of NOAA National Weather Service.

Based on the implementation of these capabilities, two main classes of WFS can be defined:

- **Basic WFS:** Implements GetCapabilities, DescribeFeatureType, and GetFeature. This is considered a read-only WFS.

- **Transactional WFS (WFS-T):** Implements all the capabilities of a basic WFS, plus the transaction operation. A transaction WFS could also implement the LockFeature operation as an option. This is considered a read/write WFS.

WFS can be used for mapping and query, as well as for geographic data clipping, projecting, zipping, and shipping. For example, the U.S. National Weather Service (NWS) Meteorological Development Laboratory provides a National Digital Forecast Database (NDFD) WFS (figure 3.14). This WFS allows the public, government agencies, and commercial enterprises to retrieve data from NDFD. It provides access to a grid forecast of weather elements such as temperature, dew point, winds, chance of precipitation, precipitation amount, weather, and sky cover.

## WCS

WCS (Web Coverage Service) supports electronic retrieval of geospatial data as "coverages," including satellite images, digital aerial photos, digital elevation data, and other phenomena represented by values at each measurement point. WCS differs from WMS in two key ways: (1) WCS returns raw data, while WMS returns a visual representation; and (2) WCS is used for raster data while WFS is generally used for vector data. WCS provides access to potentially detailed and rich sets of geospatial information, in forms that are useful for client-side rendering and input into scientific models. WCS defines the following three main operations (OGC 2006b):

- **GetCapabilities:** requests the service metadata XML

- **DescribeCoverage:** requests full descriptions of one or more coverages served by a WCS server

- **GetCoverage:** requests a coverage within the selected geographic area, within a selected range of time, in a selected coordinate system, and in a selected format such as GeoTIFF, HDF-EOS, and NITF

The U.S. National Snow and Ice Data Center (NSIDC) offers Atlas of the Cryosphere Web Coverage Services. Users can access data and information pertinent to frozen regions worldwide, including monthly climatologies of sea ice extent and concentration, snow cover extent, and snow water equivalent. These services are used as a resource in studies such as research on melting of the polar ice cap and climate change (Maurer 2007).

## OTHER WEB SERVICES

- **CSW (Catalog Service for the Web):** Catalog service is an important technology for sharing geospatial information. This Web service standard supports publishing and searching geospatial metadata. There are two types of CSW: read-only CSW and transactional CSW. Read-only CSW only supports metadata search and retrieval through such operations as GetCapabilities, DescribeRecord, GetRecords, GetRecordById, and GetDomain. Transactional CSW supports read and write of metadata, allowing users to publish, edit, and delete metadata through operations such as transaction and harvest (OGC 2007b).

  ArcGIS Server Geoportal Extension provides a platform to build geospatial portals and metadata catalogs. It exposes a CSW interface for clients to query and publish its metadata catalog (see chapter 6).

- **OpenLS (OpenGIS Location Services):** The OpenLS standard specifies interfaces that enable location-based services (LBS, see chapter 5). OpenLS includes a series of services for different usages, including route service, for determining the travel routes between two or more locations; navigation service, an enhanced version of the route service; directory service, for finding products or services; gateway service, for fetching the position of a known mobile device such as a mobile phone by its phone number; location utility service, for geocoding and reverse geocoding; and presentation service, for displaying points of interest, routes, and other information on mobile devices (OGC 2004).

- **WPS (Web Processing Service):** The WPS standard provides rules for standardizing inputs and outputs for geospatial processing services. The data required by the WPS can be delivered across a network or it can be available on the server (OGC 2007c). The main operations that WPS defines are GetCapabilities, DescribeProcess, and Execute.

- **SWE (sensor Web enablement):** The OGC SWE framework includes a series of Web service standards, including the Sensor Observation Service (SOS), Sensor Planning Service (SPS), and Sensor Alert Service (SAS). These standards enable users to discover and access the sensor data of a sensor Web or sensor network (see section 10.1.9).

## 3.4.3 RELATED STANDARDS

This section introduces GML, KML, and GeoRSS. GML and GeoRSS are data formats, while KML is a format that contains both data and visualization. Although not exactly Web service standards, these formats are often used in geospatial Web services to transport geospatial feature information between Web services and their clients.

### GML AND ITS PROFILES

**GML (Geography Markup Language)** is an ISO and OGC standard that describes how XML is used to convey geographic information. GML can describe information such as feature geographic geometry, properties, coordinate reference system, topology, and coverage (including geographic images) (OGC 2007d).

For some applications, GML is considered a complex specification, which can serve as an obstacle. To simplify its implementation and facilitate rapid adoption, OGC defined a series of GML profiles to meet specific needs. The following GML profiles are often used in geospatial Web services:

- **GML simple feature profile:** Includes points, lines, and polygons (and collections of these), with linear interpolation between vertices of lines and planar (flat) surfaces within polygons. This is often used in WFS—for example, for a GetFeature request to find all hospitals in a city, the Web service can use GML simple feature profile to return the hospitals' locations, names, specialties, and number of beds.

- **GML GeoRSS profile:** Includes geographic information in the popular RSS and Atom feeds (see section 3.4.3.3).

### KML AND KMZ

**KML (Keyhole Markup Language)** was originally created by Keyhole Inc., which was acquired by Google in 2004. The name "Keyhole" actually honors the Key Hole reconnaissance satellites, which were first launched in 1976. Google submitted KML to OGC to assure its status as an open standard, and KML became an official OGC standard in 2008.

KML is an XML-based file format for describing geographic features, including visualization. A KML file usually specifies a set of features (placemarks, images, polygons, 3D models, textual descriptions, and the like) in longitude and latitude and defines "camera view" parameters such as tilt, heading, and altitude (OGC 2008). KML also lets you designate text, pictures, movies, or links to other GIS services that appear when you click the feature. KML files are often distributed in KMZ files, which are zipped files that include KML along with its associated images and icons.

The following is a simple example of KML. It first defines a style with a certain icon, and then defines a placemark and displays it with that icon.

```
<?xml version="1.0" encoding="UTF-8"?>
<kml xmlns="http: //earth.google.com/kml/2.2">
<Document>
      <name>Placemark.kml</name>
      <Style id="sh_ylw-pushpin">
      ?<IconStyle>
      ??<scale>1.3</scale>
            <Icon>
            ???<href>http: //maps.google.com/mapfiles/kml/pushpin/ylw-pushpin.png</href>
            </Icon>
      ??<hotSpot x="20" y="2" xunits="pixels" yunits="pixels"/>
      ?</IconStyle>
      </Style>
      <Placemark>
            <name>New York</name>
            <LookAt>
                  <longitude>-73.98695099776447</longitude>
                  <latitude>40.75578596910464</latitude>
                  <altitude>0</altitude>
                  <range>1838.244767541828</range>
                  <tilt>64.84948240920286</tilt>
                  <heading>-0.00395616889357744</heading>
            </LookAt>
            <styleUrl>#sh_ylw-pushpin</styleUrl>
            <Point>
                  coordinates>-73.98695099776447, 40.75578596910464, 0</coordinates>
            </Point>
      </Placemark>
</Document>
</kml>
```

KML is often used in public information services—for example, USGS uses KML to disseminate real-time earthquake information (see figure 7.16). KML is also used in entertainment and public relations—for instance, the North American Aerospace Defense Command (NORAD), which has been tracking the path of Santa Claus every Christmas Eve since 1955, has turned to KML technology in recent years. NORAD claims that it can track Santa by detecting the infrared signature given off by Rudolph the Red-Nosed Reindeer. And so, NORAD lets kids use KML viewers to watch Santa drive his sleigh and deliver presents to children across the world. KML can also be used to display the path of a hurricane (figure 3.15).

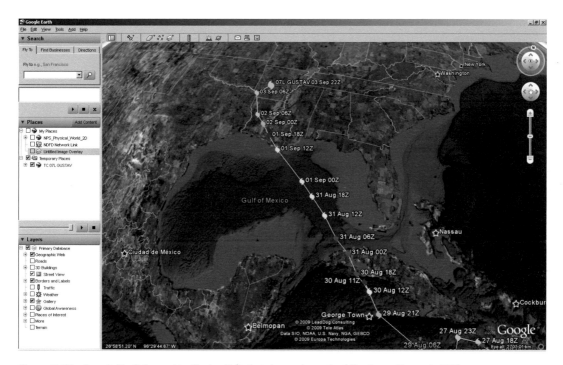

**Figure 3.15**  Google Earth is used to display KML that shows the path of Hurricane Gustav in 2008.

Courtesy of Google; LeadDog Consulting; Tele Atlas North America, Inc.; Scripps Institution of Oceanography; NOAA; U.S. Navy; NGA; GEBCOI; and Europa Technologies.

## GEORSS

Before describing GeoRSS, it is necessary to define RSS.

**RSS** (Really Simple Syndication, also referred to as Rich Site Summary, or RDF [Resource Description Framework] Site Summary) is a family of Web feed formats for publishing frequently updated information, such as newly updated blog entries and news headlines. The RSS family includes the RSS and Atom formats. The RSS format emerged in 1999 and the Atom format in 2003. Both formats have gained widespread use since 2005 and are becoming more prevalent as a way to publish and share information, including news updates.

RSS is characterized by its simplicity, its real-time status, and its feed aggregators, or feed readers. RSS formats have only a few XML tags, which describe the title, publication date, authorship abstract, and URL link to the full text for each article cited. RSS lets you subscribe to the news and your favorite Web sites as easily as clicking a mouse. RSS news readers check your subscribed feeds regularly for updates and provide a user interface so you can monitor and read the feeds at a glance, without having to check multiple Web sites.

RSS is increasingly being used for public service, by the news media and social networking sites as well as official government Web sites. CNN, the *New York Times,* Reuters, *Science* magazine, Twitter, and YouTube have all adopted this technology to spread information on selected

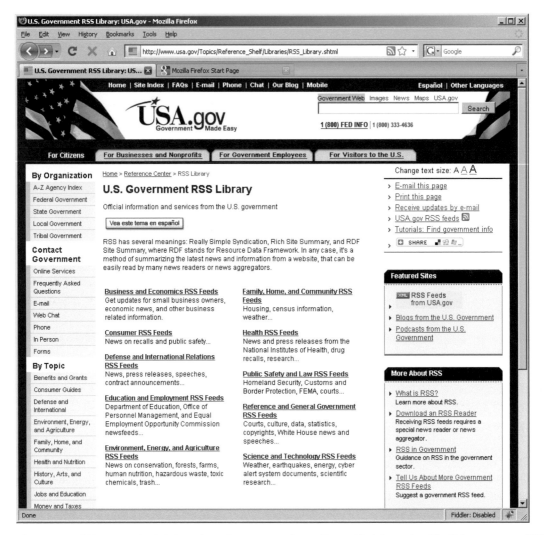

**Figure 3.16** RSS is increasingly being adopted as a means of providing public information. The U.S. government RSS library provides links to thousands of RSS feeds published by agencies in business and economics, health, public safety, science and technology, the environment, energy, agriculture, housing, census information, and weather.

Courtesy of USA.gov.

events. The World Health Organization (WHO) releases its news feeds at `http://twitter.com/statuses/user_timeline/14499829.rss` to help readers stay on top of the latest infectious diseases and contagious illnesses. The U.S. government provides thousands of different feeds cataloged at USA.gov (figure 3.16). Using RSS feeds, the Federal Emergency Management Agency publishes disaster and emergency-related news and photos, the U.S. Fire Administration publishes critical infrastructure protection information, the U.S. Census Bureau publishes current demographic survey news, the U.S. Department of Agriculture publishes economic research service news, and the U.S. Centers for Disease Control and Prevention publishes the latest disease morbidity and mortality reports.

As RSS gains greater use, people expect to see not only what is happening, but where it is happening. Location is an important aspect of RSS, and it needs to be described consistently to ensure interoperability. **GeoRSS** is a standard for encoding location information in RSS and other XMLs (OGC 2006c). GeoRSS comes in three formats: W3C Geo, OGC GeoRSS-Simple, and GeoRSS-GML. Their specifications are as follows:

- **W3C Geo:** Can only describe points in the WGS84 coordinate reference system. Though still in use, it is now obsolete.

- **GeoRSS-Simple:** Can describe basic geometries (including point, line, box, and polygon) and additional properties (including feature type, feature name, relationship tags, elevation, and radius). True to its name, GeoRSS-Simple is designed to be concise and simple. Its coordinate reference system is always WGS84 latitude/longitude.

- **GeoRSS-GML:** Supports a greater range of features than GeoRSS-Simple—notably, coordinate reference systems other than WGS84 latitude/longitude. GeoRSS-GML is an OGC GML profile (see section 3.4.3.1). If no coordinate reference system is specified, the default is still WGS84 latitude/longitude.

GeoRSS bridges the gap between nongeospatial communities and the geospatial community by extending the use of RSS. It is becoming the optimal way to request, share, and aggregate geographically tagged feeds over the Web. GeoRSS is supported by Google Maps, Bing Maps, Yahoo Maps, ArcGIS Explorer, OpenLayers, and GeoServer. GeoRSS is also commonly used in mashup applications (see chapter 4). Here are some examples of GeoRSS applications:

- Twitter provides near real-time GeoRSS feeds of who is saying what and where, allowing you to visualize the tweets on a map.

- Flickr online album provides a REST interface to return a GeoRSS feed containing a list of photos for a given area. For example, the URL `http://api.flickr.com/services/feeds/geo/United +States/California/Hollywood` returns a GeoRSS that lists photos for the city of Hollywood, which can be displayed in map viewers.

- The Global Disaster Alert and Coordination System (`http://www.gdacs.org`) provides near-real-time alerts of natural disasters, such as earthquakes, tropical cyclones, and floods, that have a potential humanitarian impact. These GeoRSS feeds are used to facilitate emergency response coordination.

- GlobalIncidentMap.com provides a series of real-time GeoRSS feeds, including Amber Alerts about abducted children, hazmat situations, threats to the White House, global terrorism, and other threats (figure 3.17).

- USGS broadcasts earthquake information in GeoRSS format (`http://earthquake.usgs.gov/ eqcenter/catalogs/eqs7day-M5.xml`). The earthquake data in California is repeatedly updated every few minutes, and for other places around the globe, it is repeatedly updated every half-hour. The GeoRSS feed implements the GeoRSS-Simple standard within an Atom format. For each earthquake, this feed uses point geometry to describe its location and a negative elevation to describe its depth (the box that follows shows a segment of this feed).

```xml
<?xml version="1.0"?>
<feed xml:base="http://earthquake.usgs.gov/" xmlns="http://www.w3.org/2005/Atom"
xmlns:georss="http://www.georss.org/georss">
     <updated>2009-05-18T18:03:24Z</updated>
     <title>USGS M5+ Earthquakes</title>
     <subtitle>Real-time, worldwide earthquake list for the past 7 days</subtitle>
     <link rel="self" href="/eqcenter/catalogs/7day-M5.xml"/>
     <link href="http://earthquake.usgs.gov/eqcenter/"/>
     <author>
             <name>U.S. Geological Survey</name>
     </author>
     <id>http://earthquake.usgs.gov/</id>
     <icon>/favicon.ico</icon>
     <entry>
             <id>urn:earthquake-usgs-gov:us:2009gua1</id>
             <title>M 5.8, near the coast of southern Peru</title>
             <updated>2009-05-18T14:01:03Z</updated>
             <link rel="alternate" type="text/html" href="/eqcenter/recenteqsww/Quakes/
us2009gua1.php"/>
             <link rel="related" type="application/cap+xml" href="/eqcenter/catalogs/cap/
us2009gua1"/>
             <summary type="html"><![CDATA[<img src="http://earthquake.usgs.gov/
images/globes/-15_-75.jpg" alt="15.655&#176;S 74.860&#176;W" align="left" hspace="20"
/><p>Monday, May 18, 2009 14:01:03 UTC<br>Monday, May 18, 2009 09:01:03 AM at epicenter</
p><p><strong>Depth</strong>: 16.00 km (9.94 mi)</p>]]></summary>
             <georss:point>-15.6552 -74.8602</georss:point>
             <georss:elev>-16000</georss:elev>
             <category label="Age" term="Past day"/>
     </entry>
     <entry>
             …
     </entry>
     <entry>
             …
     </entry>
     …
</feed>
```

### 3.4.4 THE NEED FOR EASIER STANDARDS

Easier standards such as GeoRSS and WMS understandably get wider adoption. They are supported by multiple vendors and applied in a variety of real-world applications.

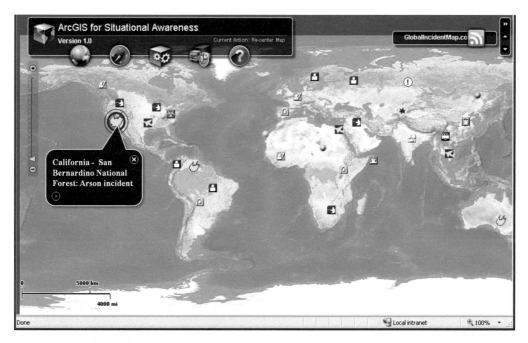

**Figure 3.17**  In this mockup, the ArcGIS for Situational Awareness Map Viewer displays a GeoRSS feed from GlobalIncidentMap.com on global terrorism and other events, including arson in San Bernardino County, California.
Courtesy of TransitSecurityReport.com, Inc., and U.S. National Park Service.

Standards also must account for complicated scenarios, in addition to simple applications, which has led to standards that are too complex to fully implement. Sam Bacharach (2006) at OGC uses an analogy of a crayon box to explain the need to simplify GML standards:

*"It's perhaps akin to requesting support for the 64 colors in the big crayon box. Did you ever watch a child with that box? Some will just pick out red, blue, green, yellow, black, orange, brown, and blue and get to work. That's enough colors. The eight selected are not overwhelming, and they offer a workable solution for drawing a picture of the family dog playing in the swimming pool in the back yard."*

While GML offers a full variety of applications, the relative simplicity of GML simple feature profile and GeoRSS profile helps widen the use of the technology, enriching the user community by reducing the complexity of GML.

Today's Web users have high expectations for Web GIS. They expect it to be as fast and as easy to use as a simple Web page. Simplified standards that speed up the process are vastly needed. Tiled map services, which use maps that are made ahead of time, are a common way to provide a quicker response and a better user experience. The Web Map Tile Service (WMTS) specification that OGC released in 2010 provides a standard-based solution to serve digital maps using predefined image tiles. Complementary to WMS standards, which are best for dynamic data or custom-style maps,

WMTS is designed to serve maps divided into tiles and often prerendered. WMTS is oriented for faster performance and higher scalability than WMS. It is highly anticipated by the geospatial community as a means of making related products interoperable and speeding up Web GIS applications to provide a better user experience.

## 3.5 OPTIMIZING WEB SERVICES

In addition to the contents and functions a Web service provides, it also can be evaluated by its quality of service (QoS). This section discusses the techniques that can achieve high Web service QoS, which is characterized by the following factors:

- **Performance:** how efficiently the system responds to users' requests, usually measured in response time

- **Scalability:** the ability to support a growing number of users without dramatically reducing performance

- **Availability:** a measure of how often a system is accessible to end users, often measured in the percentage of up time (e.g., 99.99%)

- **Security:** the ability to provide confidentiality and secure access by authenticating the parties involved, encrypting messages, and providing access control

### 3.5.1 CACHING IN ADVANCE

Caching involves generating maps and performing queries and other processes in advance rather than performing them at run time, and then storing the results for future use. If there is a cache available (figure 3.18), the Web server can quickly retrieve the result from the cache rather than ask the GIS server and database server to retrieve the data and process a user's request at run time.

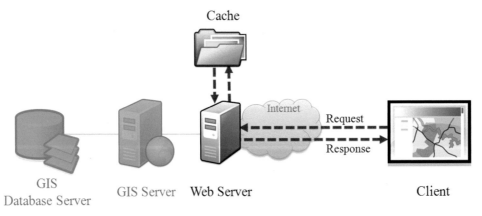

**Figure 3.18** With a cache available, the Web server can respond to a user's request quickly by retrieving the preprocessed results from the cache. This reduces the pressure on both the GIS server and the database server.

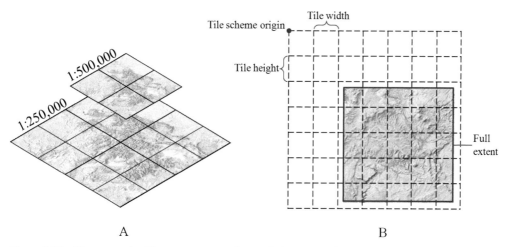

**Figure 3.19** The map cache (A) generates a set of map tile images in advance at predetermined scale levels. The map caching scheme (B) includes the number of scale levels, the scale at each level, tile dimension, tile origin, tiling area, and image format.

Courtesy of U.S. National Park Service.

Caching reduces the load on the GIS server and database server, making it an effective way to improve the QoS of Web GIS.

Caching is commonly used to generate maps and sometimes for other geoprocessing as well. **In map caching, a set of map tile images is generated in advance at predetermined scale levels for rapid display** (figure 3.19). Map caching is also called **map tiling** since each image is like a tile. The main benefits of map caching include the following:

- **Performance, scalability, and usability:** Caching reduces the load on the GIS server and the database. You can get a quicker response and get more done by having to spend less time waiting for maps to draw.

- **Cartographic quality:** Caching allows you to take the time to produce high-quality maps in advance rather than at run time.

- **Industry standard:** Consumer Web map applications make very effective use of map caching, and their popularity has changed end users' expectations of map performance. Most users expect the convenience of fast multiresolution map caches.

Before creating a cache, a number of things must be considered, such as what coordinate system to use and what tiling scheme. The tiling scheme is defined by a number of properties, including the number of scale levels, the scale at each level, tile dimension (e.g., 256 pixels wide by 256 pixels high), tile origin, tiling area, and image format (e.g., JPEG, PNG 8, PNG 24, or PNG 32). If your maps will be used with ArcGIS Online, Google Maps, or Microsoft Bing Maps, they should be in the WGS84 Web Mercator coordinate system and your tiling scheme should match theirs. Caching may take a long period of time to complete depending on the complexity of the maps and the tiling scheme, the size of the tiling area, and the scale levels involved. You can cache popular areas (e.g., urban areas) in advance and leave not so popular areas (e.g., rural areas) for **on-demand caching, where the cache is created when the area is first viewed.** A cache works best with maps that do not change frequently, such as street maps, imagery maps, terrain maps, and other basemaps. If your data tends to change, you may still be able to take advantage of caching by periodically updating the cache. You can even schedule these updates to occur automatically.

Caching can also be applied to queries and other sophisticated geoprocessing. For example, the Solar Boston Web GIS application (see chapter 2) employs caching for geoprocessing. It allows a Boston resident to see the solar potential of his or her own rooftop. Calculating solar potential is inherently computation intensive since it needs to account for a number of surrounding and atmospheric factors for individual rooftops. Solar Boston developed a model using ArcGIS Spatial Analyst to calculate the monthly solar potential, anticipated annual cost savings, and $CO_2$ savings in advance for each rooftop. The results are stored in a table. When a user wants to calculate the solar potential of his or her rooftop, the result is immediately read from the table and displayed in a chart and table (see figure 2.11). Caching hides the complexity of the geoprocessing operation and helps improve system performance and the user experience.

### 3.5.2 ALGORITHM AND SYSTEM TUNING

Software algorithms and information systems should be fine-tuned to keep them at optimum performance. This is especially true for Web applications, where a little enhanced performance can benefit a large number of users. A GIS task can be performed in several different ways. Finding the superior algorithm can greatly boost efficiency. For example, in situations where map caching is not feasible or is difficult to accomplish, dynamically drawn maps are necessary. It has been traditionally slower to produce dynamically drawn maps. ArcGIS Server improves the performance of dynamic maps by using a high-performance drawing engine and optimized map services based on MSD (Map Service Definition). With the use of optimal map services, maps can be generated dynamically and quickly and still maintain good cartographic features. GIS database tuning is also an important element of Web GIS. A few basic techniques include creating an optimal set of indexes, maintaining efficient table space partitions, defragmenting table spaces and server drives, loading data and indexes into the memory, and keeping up-to-date database statistics. A well-tuned GIS database allows the most efficient data access, which leads to better system performance and scalability.

### 3.5.3 FAILOVER AND LOAD BALANCING

Failover and load balancing are system deployment options that can increase reliability and availability through system redundancy. Failover is the capability to switch over to a redundant or standby system if an application fails or is terminated. Load balancing seeks to distribute workload evenly across two or more servers or server instances to fully use the server resources. Many large Web application systems have large server farms that can support failover and load balancing. Load balancing can be achieved through the use of hardware or software. For example, ArcGIS Server supports either the hardware or the software approach. ArcGIS Server is built with one server object manager (SOM) and one or more server object containers (SOCs). SOCs perform geoprocessing tasks while the SOM is responsible for distributing workload across the SOCs, minimizing

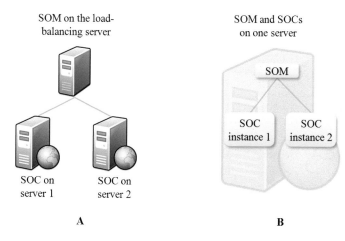

SOM on the load-balancing server

SOC on server 1     SOC on server 2

A

SOM and SOCs on one server

SOM

SOC instance 1     SOC instance 2

B

**Figure 3.20** ArcGIS Server supports load balancing through (A) the hardware approach, where multiple SOC instances run on multiple servers, as well as (B) the software approach, where multiple SOC instances run on one server.

response time and avoiding system overload. The pattern can be multiple servers with each running an SOC or one server with multiple SOCs (figure 3.20).

### 3.5.4 REDUCING THE PRESSURE ON INTERNET BANDWIDTH

Web services accept input from clients and produce output for clients, involving a significant amount of data. The data transfer between the service and the client requires sizable bandwidth. Poor bandwidth and inefficient use of bandwidth can affect the quality of Web services. To mitigate this problem, the following considerations should be made:

- **Take advantage of browser-side caching:** Browser-side caching is different from server-side caching, which generates map tiles or other results in advance. Browser-side caching involves using the content that is already downloaded to the browser rather than downloading it again from the server. This reduces pressure on the server and on the bandwidth. RESTful Web services usually have fully contained URLs, which let the client know from the HTTP header whether changes have been made to the content at a given URL. The Web client can then determine whether it is necessary to continue the download or simply to use what has already been downloaded previously.

- **Enable HTTP compression:** Compressing HTTP requests and responses on the Web server side can reduce the size of the data to be transmitted by about 50 percent, resulting in performance gains.

- **Choose the right data format:** Map picture formats come as JPEG, GIF, or PNG files, each with a different advantage. JPEGs can keep the color depth while reducing the file size of aerial and satellite imagery. GIFs and PNGs can usually preserve the sharp lines and text labels in small file sizes for vector maps. XML is a widely accepted data format for data exchange between a server and clients, but its redundant labels in start and end tags generally make it large and bulky and slow to process. In many situations, JSON, GeoJSON (a spatial data format for storing vector data based on JSON), and AMF are preferable because they are smaller and lighter.

### 3.5.5 SECURING WEB SERVICES

Web services are vulnerable to a wide range of attacks, both internal and external, because their programming interfaces are exposed to the Web over an insecure Internet. The attacks can threaten an organization's data assets, business processes, and any application that relies on Web services. Web service security is a major concern when building Web applications. A variety of techniques can be used to secure Web services, including the following:

- **Use of a private network:** The Web service and its clients are on the same private network, which is separate from the Internet.

- **Use of a VPN (virtual private network):** VPN is a technique for creating secure connections, or tunnels, across the Internet. Computers not on the LAN or VPN won't be able to communicate with the Web service.

- **Role-based authentication:** An authenticated user authorized with certain privileges can use the Web service. Certain capabilities of the Web service require higher levels of privilege.

- **Security token:** A token is an encrypted string that is derived from information about the authorized client, the date and time, or the domain of the client making the request. The token helps ensure that only authorized clients have access to the service. A token may be acquired in advance or acquired dynamically from a token service.

- **HTTPS:** HTTPS uses HTTP over an encrypted Secure Sockets Layer (SSL) connection. HTTPS encrypts the data transferred between a Web service and a client, avoiding eavesdropping and tampering.

- **Reverse proxy:** A reverse proxy server dispatches in-bound network traffic to a set of servers, presenting a single interface to the caller. A reverse proxy provides a layer of defense by separating or masking the type of server that is behind the reverse proxy.

Securing Web services is an important part of securing a Web GIS application. In a case where geospatial Web services reside on a GIS server and the GIS server is located on a private network inside a firewall (figure 3.21), clients on the Internet that are outside the firewall won't be able to directly access the GIS server. With a reverse proxy server outside the firewall, outside clients won't even know where the GIS server is located or what its IP address is. Using role-based authentication or security tokens, only authorized clients or clients with a valid token can access the Web service. Through HTTPS, all communications between the Web service and the client are encrypted, which protects against interference from tapping.

Web service technology represents a major advance in distributed computing and GIS. It is at the heart of modern Web GIS. Geospatial Web services are the driving force behind the transition from closed and isolated monolithic Web applications to open and loosely coupled Web services, which are the building blocks for mashups (see chapter 4), a basis for cloud-based GIS (see chapter 10), and the foundation for the next generation of NSDI, or NSDI 2.0 (see chapter 7). This collaborative

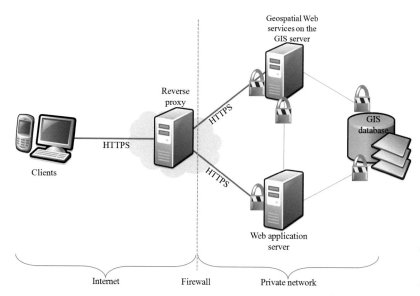

**Figure 3.21** Securing Web services is an important part of securing the whole Web GIS application. Illustrated here, the geospatial Web service is protected in the private network behind a firewall, augmented by a reverse proxy, authentication, and secured communication.

environment encourages governments and companies alike to provide their data and software as mapping, data, geoprocessing and other diverse Web services, fostering a Web services ecosystem where innovative and value-added applications will thrive. This type of mutual cooperation will maximize society's return on geospatial investment.

## Study questions

1. When did Web service technology emerge, and why is it needed?
2. What is a Web service? How does it differ from a Web page?
3. Why are Web service architecture and technology important to the geospatial community?
4. What are SOAP-based and REST-style Web services? Explain the advantages and disadvantages of each.
5. What is interoperability? How has the approach evolved in the geospatial industry? What advantages does Web-service-based interoperability provide?
6. Briefly describe WMS, WFS, WCS, GML, KML, and GeoRSS. List some products that support them.
7. What is Web Map Service, and what interfaces does it define?
8. What is GeoRSS in relation to RSS? How are they applied?
9. What are the main factors that characterize Web service quality? What are the main techniques for achieving quality?
10. What are caching and map caching in the context of optimizing Web services? Explain how they work.
11. List the main techniques to secure Web services.

## References

Bacharach, Sam. 2005. Standards drive progress in geospatial technology. *GIS Professional,* January/February.

———. 2006. The GML simple feature profile and you. *Directions Magazine,* July 18. `http://www.directionsmag.com/article.php?article_id=2224&trv=1` (accessed May 15, 2009).

Cappelaere, Pat, Linda Derezinski, John Mettraux, Keith Swenson, and Andrew Turner. 2007. RESTful OGC Services, user-centric resource-oriented architecture (ROA) approach. `http://www.slideshare.net/cappelaere/restful-ogc-services` (accessed May 11, 2009).

Dangermond, Jack. 2008. GIS and the GeoWeb. *ArcNews* 30, no. 2. `http://www.esri.com/news/arcnews/summer08articles/gis-and-geoweb.html` (accessed May 11, 2009).

EcoServ Wiki. 2010. EcoServ introduction. `http://ecoserv.pbworks.com/EcoServ-Introduction` (accessed February 17, 2010).

ESRI. 2003. Spatial data standards and GIS interoperability. ESRI white paper. `http://www.esri.com/library/whitepapers/pdfs/spatial-data-standards.pdf` (accessed May 20, 2009).

Fielding, Roy T. 2000. Architectural styles and the design of network-based software architectures. Phd diss., University of California, Irvine. `http://www.ics.uci.edu/~fielding/pubs/dissertation/top.htm` (accessed May 20, 2009).

Fu, Pinde, Al Pascual, Sarah Osborne, Keyur Shah, and Jeremy Bartley. 2008. *The best of REST: Using REST in ArcGIS Server 9.3,* ed. Carolyn Schatz. Redlands, Calif.: ESRI Press.

Greenfield, David, and Andy Dornan. 2004. Amazon: Web site to Web services. *Network Computing.* `http://www.networkcomputing.com/showArticle.jhtml?articleID=47205173` (accessed May 7, 2009).

Huang, Xiaobin. 2002. GeoAgent-based geospatial information service and application integration. PhD diss., Beijing University.

Maurer, John. 2007. *Atlas of the cryosphere.* National Snow and Ice Data Center, Boulder, Colo. `http://nsidc.org/data/atlas` (accessed May 19, 2009).

NORAD. 2007. North American Aerospace Defense Command Santa Web site going live. `http://www.norad.mil/News/2007/111907.html` (accessed May 15, 2009).

OGC. 2004. OpenGIS Location Services (OpenLS): Core services 1.0.

———. 2005. Web Feature Service implementation specification 1.1.0.

———. 2006a. OpenGIS Web Map Server implementation specification 1.3.0.

———. 2006b. Web Coverage Service (WCS) implementation specification 1.1.0.

———. 2006c. An introduction to GeoRSS: A standards-based approach for geoenabling RSS feeds, ed. Carl Reed. `http://www.opengeospatial.org/pt/06-050r3` (accessed May 6, 2009).

———. 2007a. Glossary of terms—I. `http://www.opengeospatial.org/ogc/glossary/i` (accessed May 12, 2009).

———. 2007b. Web Catalog Service Specification 2.0.2.

———. 2007c. Web Processing Service 1.0.0.

———. 2007d. OpenGIS Geography Markup Language (GML) encoding standard 3.2.1, ed. Clemens Portele.

———. 2008. KML 2.2.0.

Richardson, Leonard, and Sam Ruby. 2007. *RESTful Web Services.* Sebastopol, Calif.: O'Reilly & Associates.

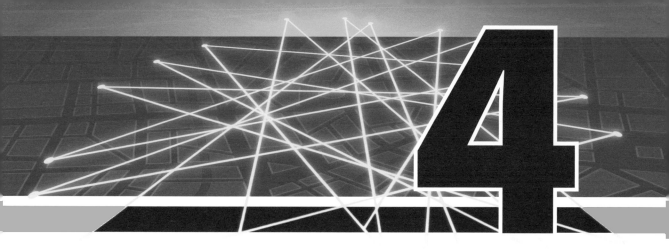

# GEOSPATIAL MASHUPS

The mashup, a hallmark of Web 2.0, is a genre of interactive Web applications that seamlessly integrate content retrieved from diverse Web-based resources to create innovative applications. While the concept has existed since the late 1990s when Web services originated, mashups have become more popular since 2005, largely as a result of browser-side APIs. Browser-side APIs have greatly reduced the complexity of integrating multiple Web resources, making mashups easier.

The mashup phenomenon provided an avenue for research. Mashups are a sign of progress in many areas of geospatial science and technology. Integration of multiple data layers or datasets, often from multiple sources, is one of the most common functional requirements of GIS applications. With the use of browser-side APIs, mashups demonstrate an easy method to accomplish this data integration. The simplicity of building mashups has democratized GIS application development and extended the use of GIS to neogeography applications. Providing an easy way to use the Web resources available on other people's computers, including others' GIS data and software capabilities, mashups also are the impetus for cloud-based GIS, or GIS on the Web, and the next generation of spatial information infrastructure.

Mashups still face challenges in many respects. It can be difficult, for instance, when it comes to integrating unstructured contents such as HTMLs. Yet a huge amount of information that is invaluable to a potentially unlimited range of applications lies hidden within billions of Web pages. Extracting, georeferencing, and using this information to produce new stores of information is an important research area for geospatial data mining and mashups.

This chapter introduces the mashup concept, evolution, and impact. It then categorizes the main Web resources and summarizes the characteristics of their programming interface. To further illustrate the mashup design pattern, the chapter describes a real-world mashup application for emergency response. Best practices are presented at the end of the chapter, along with prospects for the use of mashups.

## 4.1 EVOLUTION AND IMPACT

The etymology of the term "mashup" derives from pop music, with musicians creating new songs by remixing, or mashing up, multiple sound tracks. Mashups gained a new meaning in the Web application context in 2005 with the emergence of browser-side APIs, attracting a surge of interest from both professional and casual GIS users, though the idea started to grow in the late 1990s.

### 4.1.1 REMIXING THE WEB

The content explosion over the WWW provided both the foundation, and the need, for mashups. The appeal of the Web has encouraged the participation of government, businesses, institutions, and a vast array of Internet users. Every second, maps, Web services, Web pages, blogs, photos, and videos are being added to the Web or are being updated. Yet the way such contents and functions are organized or presented doesn't always meet users' needs. A user often needs information from more than one Web site, which fuels the need to combine the contents of multiple Web sites into a new application, or mashup application (figure 4.1).

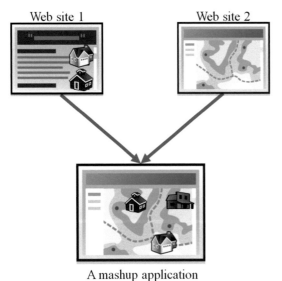

A mashup application

**Figure 4.1** Mashup applications are created by combining the contents and functions of multiple Web sites into a new application.

The following applications are examples of mashups (figure 4.2):

- Web sites such as BizRate and PriceGrabber extract commodity prices from different retail sites, and then present these prices on one page, allowing Web shoppers to find the best deal at a one-stop site rather than visiting multiple sites.

- The HousingMaps Web site (`http://www.housingmaps.com`) combines housing information from Craigslist and Google Maps, so that users can see at a glance where houses for sale are located.

- The Zillow Web site (`http://www.zillow.com`) maps estimates of home values and displays their price, associated facts, taxes, and school zones on top of Microsoft Bing Maps.

- The CrimeMapping Web site (`http://www.crimemapping.com`) displays crime data on top of ArcGIS Online maps to provide the public with information on crime activity by neighborhood.

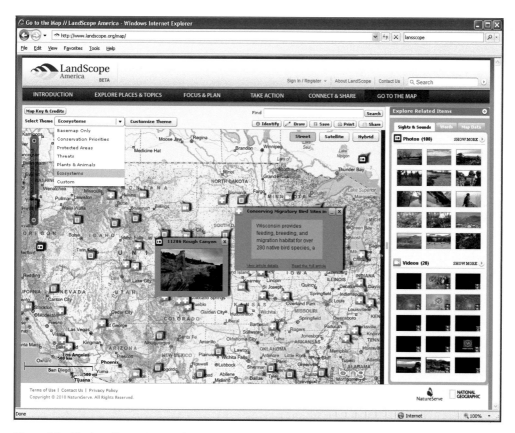

**Figure 4.2** The LandScope America (`http://www.landscope.org/map`) Web mapping application combines conservation map layers from NatureServe servers, a rich set of multimedia data from MetaLens, and Microsoft Bing Maps to present a new mashup application that provides an online resource for nature preservation.

Courtesy of NatureServe, Microsoft, and National Geographic.

- The LandScope America Web site (`http://www.landscope.org/map`), developed by NatureServe and National Geographic, combines conservation-related maps, such as conservation priorities, protected areas, threats, and ecosystems, with Microsoft Bing Maps. The Web site exhibits a rich site of photos, audio clips, and video clips about plants, animals, and other nature features on top of these maps.

- Oregon State Stimulus Tracker (`http://www.oregon.gov/recovery/StimulusReporting`) combines information about where national stimulus dollars are being spent with Web maps from ArcGIS Online. It keeps citizen informed on where the stimulus money is headed and the share that each sector of the economy is receiving.

- The Twitter Map Web site (`http://maps.esri.com/Labs4/twittermap`) retrieves people's tweets from the Twitter Web site and displays the tweets on top of basemaps from ArcGIS Online so that users can see what people are talking about at a particular location.

- Most user-created maps at ArcGIS.com combine ArcGIS Online basemaps with geospatial Web services hosted by various contributors.

From these examples, a simple definition can be derived:

**A mashup is a Web page or application that dynamically combines contents or functions from multiple Web sites.**

This definition includes three fundamental points:

- **Web contents or functions:** These are the Web resources used to build mashups. They can be contents (such as data and maps) as well as functions (such as operations or processes). For example, analytical Web services often encapsulate certain types of business logic, where the user sends input parameters, and the service returns the results from using the logic. In this case, it is functions rather than just contents that are being combined.

- **Dynamically:** This means that the mashup is generated automatically through computer programs. A mashup maintains certain levels of live linkage to the sources it uses. If the source is updated, the mashup application should automatically update within a certain time frame. If you manually capture a screen image of a map, save it as a JPEG file, upload it to your Web site, and overlay information on top of it, the Web site is not considered a mashup, since it has no dynamic link to the original source. (If this type of scenario was considered a mashup, it would mean that any Web site that cited a sentence from another Web site would also be considered a mashup).

- **Multiple Web sites:** One of these sites can be from the mashup application itself. If your Web site overlays your pictures on a map from another Web site, your Web site is a mashup.

While mashups are applicable to many disciplines, this chapter focuses on "geomashups." Given the preceding definition for mashups, geomashups can be similarly defined:

**A geomashup is a mashup where at least one of the contents or functions is georeferenced.**

Geospatial mashups integrate multiple data sources based on common geographic locations. Mashups clearly carry overtones of the overlay function. Overlay, as represented by the iconic "layer cake" view of GIS that appears on the cover of textbooks and in the logos of GIS centers (see figure 1.4), has been a central concept in the development of GIS since its inception (Goodchild et al.

2010). Overlay includes topological overlay, which restructures multiple data layers into one vector dataset, and graphic overlay, which involves superimposing visual images one on top of the other.

Today's mashups are mostly graphic overlays that leave users with the task of drawing inferences from the visual display. A topological overlay, however, is needed and can be implemented for certain mashup applications—for example, overlay flood boundaries with city boundaries to calculate the areas flooded in each city, or overlay the flood boundaries with a parcel layer to find out whose properties need to be inspected for flood damage. Overlays traditionally require that the various data layers be stored on the same local computer. Mashups are able to retrieve distributed data from the Web and dynamically overlay these layers.

## 4.1.2 SERVER-SIDE VERSUS BROWSER-SIDE MASHUPS

The concept of mashups can be tied to the emergence of Web services. The original intent of Web services research and development was to use Web services as programming components or building blocks, so that people could combine multiple Web services, often on multiple Web servers, to build new applications. For example, ESRI Geography Network emerged in 2000 with a Web map viewer called ArcWeb Explorer, which allowed users to dynamically choose and merge map services from multiple ArcIMS Web services. For instance, a toxic release map service from the U.S. Environmental Protection Agency could be integrated with a basemap service from the U.S. Geological Survey to find the potential damage from a toxic spill.

Prior to 2005, mashups were a top-down effort advocated by major software vendors, government, and academic institutes. Mashups relied on what Web services could offer, and the mashup processes were done on the server (figure 4.3). The advantage of such mashups was that the server had more powerful hardware and software than the browser, but these mashups also required professional programming tools and complex server-side programming, which was labor-intensive when it came to development and deployment. Because of this complexity, the original mashups were mainly limited to professional programmers. And because the Web services at the time were not

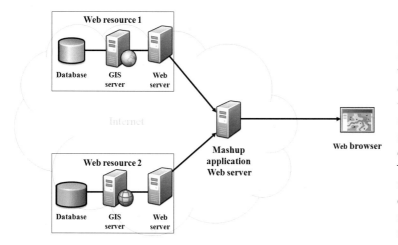

**Figure 4.3**   In the server-side mashup architecture, the mashup Web server sends requests to different Web services, receives the responses, and merges the results. This requires server-side programming, which is usually done by professional developers. This was the main form of mashups prior to the emergence of browser-side Web APIs and remains an important form of mashups in certain situations.

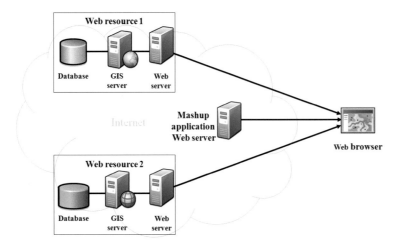

**Figure 4.4** In the browser-side mashup architecture, the Web browser sends requests to different services, receives the responses, and displays the composite results. This type of mashup, which typically uses lightweight programming and is easy to develop, has become the most popular mashup format since 2005.

as rich as they are now, mashups did not gain immediate popularity. Nevertheless, the concept and the technology laid the foundation for the subsequent flourish of geomashups.

Today's mashups are mostly accomplished on the browser side (figure 4.4). Browser-side mashups sprang from the grass roots. Their wide and rapid adoption since 2005 stemmed from the rich contents and services available over the Web and the popularity of AJAX (asynchronous JavaScript and XML, see chapter 2) and JavaScript. Many consumer Web mapping sites provide maps via JavaScript, which is so simple that many Web users can learn it without professional training. Web users can also view the sources of the scripts using the browser's viewer source function. Thus, developers could crack the code and build applications from it—for instance, marking houses on a map. Realizing the number of users that mashup applications could attract and the advertising potential such a large number of users could have, consumer mapping companies such as Google, Microsoft, Yahoo, and MapQuest officially released their mapping capabilities via the JavaScript API. Professional GIS companies also adopted this approach as an easy and quick way to develop Web applications. For example, ESRI released ArcGIS APIs for JavaScript, Flex, and Silverlight. These APIs greatly simplify the user experience for users and developers and reduce the effort needed to build critical and enterprise-level applications with mashups. The simplicity, vendor support, and public interest in using mashups have all led to the increasing popularity of the genre.

### 4.1.3 SIGNIFICANT IMPACTS ON GIS

Integration of multiple data layers or datasets, often from multiple sources, is one of the most common functional requirements of GIS applications. Recent advances in mashup technology, which facilitate the integration of distributed datasets, therefore produce many impacts on GIS.

## NEW DESIGN PATTERN FOR WEB GIS APPLICATIONS

The plethora of mashups demonstrates a common design pattern that consists of basemaps, operational layers, and tools (see equation 2.1)

- **Basemaps** provide a geographic frame of reference—for example, the street maps or aerial imagery maps used in the examples earlier in this chapter. The basemaps are often from the Web sites or Web services of Google, Microsoft, ArcGIS Online, Yahoo, or MapQuest.

- **Operational layers** (or thematic layers) show items of interest on top of a basemap—for example, the houses, crime sites, nature spots, stimulus dollars, and tweets in the earlier examples. They are often from Web sites or Web services that are different from the basemaps. Operational layers usually respond to user actions such as a mouse click by displaying the requested information in a pop-up window.

- **Tools** can execute certain business logic or analytical functions—for example, the find address and find place functions in the preceding mashup examples. Enterprise mashups usually can perform more specialized or advanced tasks.

These components are often integrated with the use of browser-side APIs. This pattern provides a quick and simple way to build Web GIS applications (Brown 2009; Brown et al. 2008). It has been widely adopted by both professionals and amateurs (see section 2.3.3), aided by modern server GIS products, which enable organizations to publish their contents and functions as Web services, especially REST Web services. Today's Web GIS applications are mostly mashups that are developed with this pattern.

## DEMOCRATIZATION OF WEB GIS APPLICATION DEVELOPMENT

Web GIS marked a milestone because it expanded the range of GIS users from GIS professionals to the general public. The mashup approach marks another milestone. It expanded Web GIS application development from GIS professionals to anyone with basic programming skills. The simplicity of the mashup has encouraged public participation and provided a platform for people to unleash their creativity. There are mashup applications mapping family photos, secret fishing holes, celebrities, basketball tourneys, the Olympics torch relay, auctions, biking trails, Web cameras, tweets, and news events. While many mashups don't have a clear business model and are done for fun, they demonstrate that GIS can be applied in much broader areas than the traditional technical fields.

Mashups extend GIS to neogeography applications, which are used for personal and community activities, often by nonexpert users. The mashup phenomenon encourages user-generated content (UGC) and volunteered geographic information (VGI, see section 10.1). The diverse content, fast performance, and rich user experience provided by mashup applications appeal to a large volume of users, promoting the development of public participation GIS (PPGIS, see section 10.2).

## THE WEB AS BOTH DATABASE AND GIS SERVER

The phrase "the network is the computer," often attributed to John Gage of Sun Microsystems, imagines a future in which the ability to compute extends freely beyond your own machine to include services distributed across the entire Internet. We are still a ways off from achieving this ideal, but mashups demonstrate the possibilities in the context of GIS (Goodchild et al. 2010). Organizations that can't afford or don't possess their own database, GIS server, and the necessary expertise can use the mashup architecture to get the data, maps, and models they need from the Web.

Mashups can potentially combine any type of contents and functions over the Web, regardless of whether a formal programming interface is available. The vast majority of contents over the Web are HTML pages and photos that don't have formal APIs, but they contain a huge amount of valuable information that can be scraped and geospatially tagged or referenced (see section 4.2.3), and then reused to build new and value-added applications. Virtually the whole Web can be remixed, which opens the door to unlimited value-added mashup applications.

Mashups are an interesting phenomenon in science and technology that represents the essential characteristics of Web 2.0—that is, simple and lightweight programming, use of the Web as a programming platform, and the bottom-up flow of information from Web users who combine their own data with that of other sources to build fascinating Web applications.

## 4.2 WEB CONTENTS, FUNCTIONS, AND INTERFACES

Web contents and functions are the foundation of building mashups. These Web resources can be classified into two major categories (figure 4.5)—with Web APIs or without APIs. Web resources with Web APIs include Web services and browser-side APIs. These resources, which primarily consist of basemaps, other theme maps, and commonly used functions, are easy to mash up. Web resources without Web APIs are mostly HTMLs. While these resources can be difficult to use in mashups because of the complexity of data extraction needed, they have a wealth of information and hold great potential for use in countless mashups.

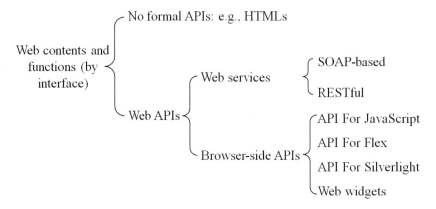

**Figure 4.5** Web contents and functions are classified by their programming interface.

### 4.2.1 BROWSER-SIDE APIS

Mashups are easy when they use Web resources that can be accessed with browser-side APIs. Browser-side APIs are easy to consume and are by far the most popular type of API used in mashups. They have several forms, including the JavaScript API, Flex API, Silverlight API, and Web widgets. JavaScript (figure 4.6) is the most widely used programming language over the Web. With the addition of toolkits like Dojo, JavaScript can accommodate sophisticated graphics and user interactions. The Flex and Silverlight APIs are more expressive than JavaScript; have excellent graphics support, animation effects, and flexible user interaction capabilities; can provide a good user experience; and are especially suited for developing rich Internet applications (RIAs, see chapter 2).

Browser-side APIs are easier to use than Web services because they let users leverage these services without knowing what goes on behind the scenes. Browser-side APIs handle Web services presentation as well as user interactions on the browser (table 4.1).

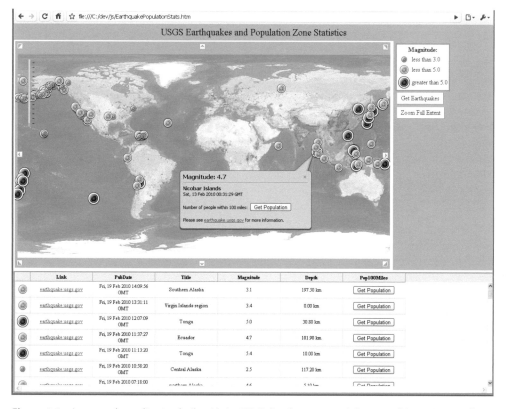

**Figure 4.6**   An example application built with ArcGIS Online basemap and demographic services via the ArcGIS API for JavaScript displays the current earthquake magnitudes and locations from USGS and then calculates the number of people within 100 miles of an earthquake.

Courtesy of U.S. National Park Service and U.S. Geological Survey.

| | **WEB SERVICES** | **BROWSER-SIDE APIS** |
|---|---|---|
| **EXECUTION LOCATION** | Web servers | Web browsers |
| **LANGUAGES TO USE** | Independent of programming languages. Can be used with any language. | Dependent on specific programming languages. APIs for JavaScript, Flex, and Silverlight need to be used with corresponding language. |
| **CAPABILITIES** | Provide server-side functions, can operate databases. Do not provide user interface or user interaction functions. | Provide browser-side functions, including user interface, user interaction, mouse control, browser-side logic, etc. |
| **RELATIONSHIP** | Waiting to be called by browser-side API or other languages. | Usually encapsulate Web services, especially RESTful Web services. Can call Web services based on user interaction, and let users use Web services without knowing it. |

**Table 4.1**   Comparison of Web services and browser-side APIs

Browser-side APIs, usually built on top of Web services, allow developers to build Web applications faster and easier. Web services only perform server-side functions (figure 4.7), and so usually leave much of the application development work to the browser side—for example, map display, mouse control, map parameter calculation, sending requests, receiving responses, and parsing results. Browser-side APIs, on the other hand, can do all these jobs for you and hide the interactions with Web services so that you won't even notice. To use a metaphor, if Web services are like chefs, browser-side APIs are like waiters and waitresses bringing you the finished dish. Many organizations are providing their Web resources via browser-side APIs. Well-known ones include the street maps, aerial imagery, address and place search, and routing functions provided by Google, Microsoft, Yahoo, and MapQuest. ArcGIS Online provides a rich set of business analysis functions (see chapter 8). ArcGIS.com has a portal that allows ArcGIS communities to share their Web services (see chapter 7). These services can be accessed via ArcGIS Server APIs for JavaScript, Flex, and Silverlight.

The ArcGIS APIs for JavaScript, Flex, and Silverlight demonstrate the characteristics of browser-side APIs. In addition to providing user interaction-related functions, these ArcGIS Web APIs consume ArcGIS Server REST Web services, which can deliver mapping, imagery, query, feature editing, tracking, geometry operations, routing, and more advanced geoprocessing tasks (see details at **http://resources.esri.com/arcgisserver/index.cfm?fa=applications**). These APIs enable you to do the following tasks easily and quickly:

- Embed an interactive map of your own data or maps from your ArcGIS server, ArcGIS Online, Google Maps, or Microsoft Bing Maps in your application

- Display your operational data as graphics to support rich user interactions and reports

- Search for features or attributes in your GIS data and display the results

- Edit feature geometry and attributes

- Locate addresses and places

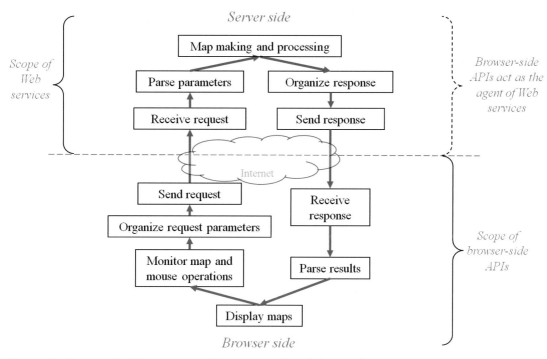

**Figure 4.7** Browser-side APIs encapsulate Web services and handle browser functions, making Web GIS application development faster and easier for users.

- Request servers to execute custom GIS models on the server side, and receive and display the results on the browser side

ArcGIS browser-side APIs support the mashup design pattern of basemaps, operational layers, and tools by providing the corresponding programming components:

- **Maps:** Corresponding to the basemap and sometimes to the operational layers. These maps can display different types of map services, including ArcGIS Server map services (both cached and tiled), ArcIMS map services, ArcGIS Online maps, Google Maps, Microsoft Bing Maps, WMS, and KMLs. The maps have basic navigation, mouse, and keyboard event listeners built in; thus, they can interact with users without additional programming.

- **Graphics:** Typically used to represent the operational layers, such as observations, sensor feeds, incidents, and results of query and analysis (see equation 4.1).

**EQUATION 4.1**

```
Graphics = Geometry + Attributes + Symbols + Info Windows
```

Graphics are vector features (points, lines, and polygons along with their attributes) downloaded to the browser side. They are rendered on the fly, usually based on a single attribute—for example, flying helicopters can be displayed with animation icons and earthquakes can be classified by magnitude—or by multiple attributes (for example, states can be displayed with pie charts that show the percentage of the population by race.) Graphics can also show spatial distribution—for example, points close to each other can be displayed in clusters, and the density of points can be displayed by a heat map (figure 4.8). Graphics can improve user interaction. Hovering above or clicking the graphics can produce information windows to show attributes, charts, pictures, audios, videos, or HTML links for further details.

- **Tasks:** Interact with the back-end ArcGIS Server to perform query, identification, editing, geocoding, routing, geometry processing, and other geoprocessing tasks (figure 4.9).

Web widgets further simplify mashup development. Web widgets, including Dijits (i.e., Dojo widgets), code snippets, and other forms, are portable segments of code or objects that can be easily embedded within a Web page. Web widgets are often built on top of browser-side APIs. They usually package several lines of HTML and JavaScript or embed a Flex or Microsoft Silverlight object on a Web page. Web widgets can communicate with other Web sites, perform certain types of logic, and bring other contents and functions to the Web page where the widget is embedded (figure 4.10).

### 4.2.2 WEB SERVICES BEHIND THE SCENES

In the mashup world, it's the browser-side API that holds much of the glory, but a lot of its strength comes from the Web services behind the API. Web services, the programs that run on a Web server and expose programming interfaces to the Web, are often used in mashup applications. RESTful Web services primarily rely on simple URLs and are easy for browser-side programming languages to request and process in a mashup. The browser-side APIs introduced earlier typically wrap RESTful Web services. While SOAP-based services are not as easy for browser-side programs to consume as RESTful Web services, SOAP Web services have solid support in server-side, desktop, and mobile application development using technologies including Java and .NET.

Web services have related standards, such as OGC standards for WMS, WFS, WCS, GeoRSS, and KML (see chapter 3). The simplicity of GeoRSS and the popularity of KML make them widely used formats that can be used without programming. These standard services are supported by many geobrowsers, including ArcGIS Explorer, NASA World Wind, Google Earth, and a variety of applications developed using ArcGIS APIs for JavaScript, Flex, and Silverlight.

If the resources that are needed are not readily available online, organizations can publish their own Web services using their own data. Government agencies, including USGS, the National Oceanic and Atmospheric Administration, U.S. Environmental Protection Agency, and many others, are among the advocates of Web service initiatives that publish useful services. Many examples of Web services have been given in chapters 3 and 7. More government Web services can be discovered through Web portals such as geodata.gov (`http://www.geodata.gov`). Commercial companies are also important Web service providers (see chapters 7 and 8). For example, GeoNames provides a number of address, place, and administrative unit search functions through its Web services. Its RSS to GeoRSS encoding Web service is unique and useful for mashups. RSS feeds

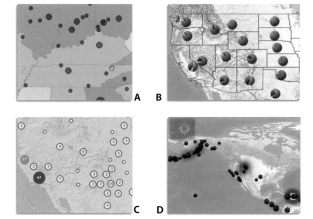

**Figure 4.8** The ArcGIS Server APIs for JavaScript, Flex, and Silverlight can render graphics on the browser side in a number of ways. For example, (A) points can be displayed in different sizes corresponding to a certain attribute; (B) pie charts can be used to show state populations by the percentage of race; (C) points close to each other can be displayed in clusters; and (D) points can be displayed in a heat map to show the density of the points.

Courtesy of U.S. National Park Service, U.S. Geological Survey, and U.S. Census Bureau.

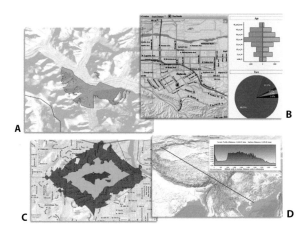

**Figure 4.9** The ArcGIS Server APIs for JavaScript, Flex, and Silverlight can call ArcGIS Server to perform query and geoprocessing. For example, (A) a viewshed analysis model to calculate the visible areas from a mountainous location; (B) a spatial query to get census block demographic information, such as age and race; (C) a service area analysis to calculate the service area of a fire station or retail store (calculated by driving distances within 1, 2, and 3 minutes); and (D) a terrain profile analysis to obtain the ground surface elevation change along a line.

Map data ©AND Automotive Navigation Data; courtesy of Tele Atlas North America, Inc., and U.S. National Park Service.

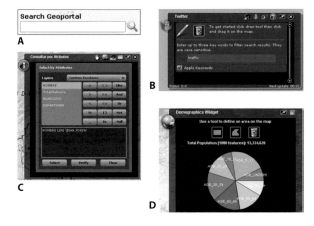

**Figure 4.10** Examples of Web widgets include (A) an ArcGIS Server Geoportal search widget, which can be embedded in an HTML page; (B) a Twitter feed widget that can retrieve and display tweets for a given region; (C) a query builder widget that can build query expressions based on the attributes in a Web service data layer; and (D) a demographic widget that can retrieve and chart the population information for a given region.

© Courtesy of Twitter. (D) Data courtesy of U.S. Census Bureau.

(including RSS and Atom, see chapter 3) usually include real-time information but are not necessarily spatially referenced, so they cannot be directly mapped. GeoNames has an RSS to GeoRSS conversion server that can extract the address and place-names from RSS, find latitudes and longitudes, and inject the coordinates into GeoRSS. For example, the U.S. Centers for Disease Control and Prevention has an RSS feed that reports recent outbreaks and incidents of illness at `http://www2a` `.cdc.gov/podcasts/createrss.asp?c=233`. To see where an outbreak or incident occurred, you can simply pass the news RSS URL to GeoNames in a URL like `http://ws.geonames.org/` `rssToGeoRSS?feedUrl=http%3A//www2a.cdc.gov/podcasts/createrss.asp%3Fc%3D233` (which is a URL encoded version of `http://ws.geonames.org/rssToGeoRSS?feedUrl=http%3A//www2a.cdc.gov/` `podcasts/createrss.asp%3Fc%3D233`). The result will be returned in GeoRSS format, which can be used in many mashups.

### 4.2.3 WEB RESOURCES WITHOUT A FORMAL INTERFACE

The majority of resources over the Web are Web pages (HTMLs) that have no formal interface (i.e., API). Web pages frequently contain a wealth of useful data that can fulfill a variety of real application needs, even without a formal API. The data can be extracted, spatially referenced, and integrated with other important applications (figure 4.11).

For example, the World Health Organization Web site lists chronologically the areas with local transmissions of SARS in 2003 (figure 4.12), H1N1 virus (swine flu) in 2009 and 2010 (`http://www.` `who.int/csr/disease/swineflu/updates/en/index.html`), and other infectious diseases. While this

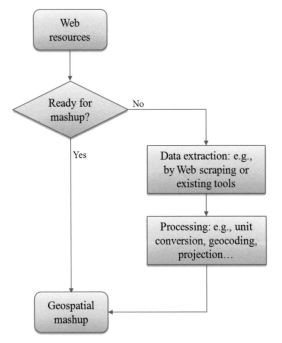

**Figure 4.11** Web resources without a formal API are not ready for a direct mashup. Useful data needs to be extracted using Web scraping technology or other tools. The extracted data often needs to be georeferenced, using geocoding or coordinate system conversion, to be used in a geospatial mashup.

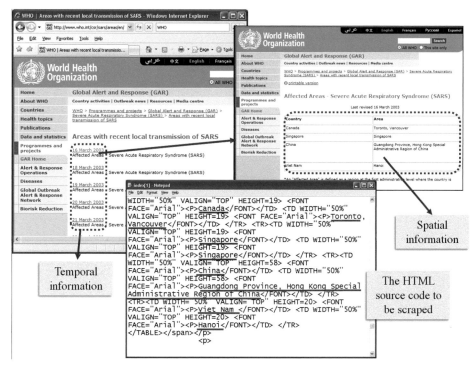

**Figure 4.12** The World Health Organization Web site (`http://www.who.int/csr/sars/areas/en/`) lists chronologically the areas with local transmission of SARS in 2003. On the upper left, this HTML page has temporal information. On the upper right, this HTML page has spatial information. At the bottom, the HTML source code that needs to be scraped has the place-names underlined that need to be extracted. All the information is valuable for mashup applications for mapping and analyzing the spatial and temporal patterns of SARS transmission.

Courtesy of World Health Organization.

content doesn't have a formal programming interface, it can be useful for mapping and analyzing the spatial and temporal transmission patterns of infectious diseases.

Web pages are usually a mixture of contents and styles (such as font size, color, and page layout) that are designed for reading. To be used in a mashup application, the useful content needs to be separated from the presentation, and then extracted and georeferenced. This is accomplished with Web scraping and sometimes geoparsing techniques (see section 10.1.5). Screen scraping is a technology in which a computer program extracts data from the display output of another program. As a new breed of screen scraping, **Web scraping, also referred to as Internet screen scraping or Web data extraction, extracts information from unstructured Web content, typically in HTML format, and transforms it into structured data.** In essence, with the Web scraping technique, developers need to carefully analyze the HTML source code of the Web page, find out the string patterns related to the useful information, write specific code to process the strings, and extract the information, which can then be used in a mashup.

If the contents extracted by Web scraping do not contain georeferences—either geographic coordinates (e.g., latitude and longitude) or a key (e.g., name or ID) that can be linked to other geo-referenced contents—the contents will need to be further processed, usually through geocoding or by using the place finder function. This process of parsing content, extracting geographic references, and resolving the geographic meaning is a form of **geoparsing** (see section 10.1.5).

Web scraping and geoparsing usually need to be done on the server side for two reasons: (1) these processes are too complex for lightweight programming and usually require more powerful languages such as server-side .NET or Java; and (2) the server does the processing one time so that all browsers can use the results. The output of Web scraping and geoparsing should be in XML format (e.g., GeoRSS), JSON, or other structured formats, for consumption in mashup programs.

Web scraping requires a significant amount of programming effort, and the scraping program for one Web site generally doesn't work for another Web site. To address these problems, Dapper (`http://www.dapper.net`) and other tools provide the ability to visually match the Web content to a particular structure, and extract and convert the content to a choice of structured formats such as RSS. While these tools only work for a limited variety of HTMLs, they represent a step forward in automating the content extraction from unstructured Web resources.

## 4.3 MASHUP DESIGN AND IMPLEMENTATION

Many of today's Web GIS applications are mashups, as they involve contents and functions from multiple Web services or multiple Web resources. Understanding the steps to building mashups is helpful for the design and implementation of Web applications in general.

### 4.3.1 STEPS TO BUILDING A MASHUP

Several considerations must be made before building a mashup application. You should generally take the following steps:

1. Identify the problems to solve and define the objectives to accomplish.

2. Search for the Web resources you need by using geoportals (see chapter 6) or Web portals. Evaluate each resource by its quality and usefulness per the project requirements and by its programming interface and the workload required to mash it up.

3. Publish your contents and functions when they are not available over the Web, and when yours are better for the project at hand.

4. Organize the contents and functions into basemaps, operational layers, and tools.

5. Implement your application through configuration or programming:

   - To configure: Since existing geobrowsers and map viewers can mash up a variety of sources without programming, you may just need to change some configurations. With Web GIS platforms such as ArcGIS.com, users can search for various geospatial contents contributed by the ArcGIS.com user community and interactively create mashup applications without the use of programming.

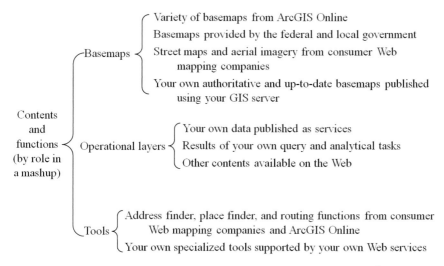

**Figure 4.13** Building a mashup often involves discovering available Web resources, publishing your own Web resources, and then seamlessly integrating them into a purposeful mashup.

- To program: Based on the programming interfaces of the contents and functions you chose, you may need to write programs from scratch or use existing frameworks, such as those available from ArcGIS Online communities `http://resources.esri.com/arcgisserver/ index.cfm?fa=community`.

6. Test, enhance, and deploy the application.

A mashup application depends on Web resources. Recent advances in Web services and cloud computing—that is, services provided by highly scalable computer centers on the Web—have led to an increased number and variety of Web services that are available online. For many applications, however, you will often find that the Web resources you need are not directly consumable, or not even available over the Web. In such cases, the resources need to be preprocessed (including data extraction, reformatting, reprojection, and other processes), or you'll need to create them yourself (figure 4.13). Each organization has its own authoritative, up-to-date, and sometimes mission-critical data and models. A city, for example, might have its own aerial imagery that is more recent and of higher resolution than what is available from consumer Web mapping sites. A company usually has highly accurate underground pipeline data, critical infrastructure, private helicopter locations, customer addresses, and business analyst models. The city and the company can publish these data, maps, and models as Web services, secure them if needed, and then use them in mashups to build applications that fulfill specific business needs.

## 4.3.2 CASE STUDY: A MASHUP FOR SITUATIONAL AWARENESS

Mashups, while suitable for a variety of applications, are especially fitting for situational awareness, emergency response, and use in an executive dashboard. They are useful for applications where managers and executives need accurate, complete, and timely information for sound decision making. Mashups are all about combining dynamic sets of information to provide a complete view of a situation. Mashups aggregate various sources of information with the source data left in the hands of its custodian. That way, when the source is updated, a mashup can show the updates in real time. These characteristics allow decision makers to gain accurate situational awareness as the events unfold so they can make the right decisions quickly.

AEGIS (Advanced Emergency GIS) at Loma Linda University Medical Center (LLUMC) in Loma Linda, California, represents a successful example of using mashups for emergency medical response. AEGIS is a Web-based hospital situational awareness system that monitors and maps the locations and status of emergency management-related resources—including area hospitals, ambulances, rescue helicopters, and other field responders—covering a broad regional area. The magnitude of a medical emergency is usually not as extensive as an earthquake, wildfire, or tsunami, but medical emergencies happen every day, and the medical emergency response needs to operate well every day. The approach demonstrated in this mashup is useful for a variety of emergency response applications.

LLUMC is located in the Inland Empire region of Southern California. It encompasses a vast geographic area, and with nearly 4 million residents, it is one of the fastest growing regions in the United States. It is also one of the most prone to earthquakes and wildfires. As the largest hospital in the region, LLUMC has a responsibility to be prepared to respond to emergencies of every nature. In 2005, LLUMC partnered with ESRI to develop a Web-based emergency response system that would coordinate all emergency resources and be accessible to the community.

### DEFINING PROJECT OBJECTIVES

The goal of AEGIS was to develop an automated visualization and decision support tool for the hospital's Department of Emergency Medicine. It needed to be a Web-based situational-awareness GIS that would monitor and map the location and status of emergencies, locate victims and emergency response personnel, and track other factors that could impact emergency response. The project would integrate a variety of technologies into a scalable visualization model, providing a coordinated hub to maximize and leverage all health-care resources available at the base hospital. AEGIS would facilitate assessment, handling, and tracking of patients at the site of an incident (e.g., automobile accident, fire, cardiac arrest, violent crime, earthquake, or other health-care emergency) as well as during transportation of the patient to an appropriate care facility (Fike 2007; Kolbasuk McGee 2008).

Version 1 of this project was developed in 2005 with ArcIMS. Version 2 was done in 2008 with ArcGIS Server. Version 2 takes advantage of cached map services and the expanded capabilities of the ArcGIS API for JavaScript and ArcGIS Mobile. The ArcGIS API for JavaScript supports easier and more expressive mashups, while ArcGIS Mobile allows implementation of mashups on mobile devices to facilitate communication and collaboration between the dispatch center and field personnel. This section discusses a subset of this application in the mashup context. The mobile component was developed using ArcGIS Mobile (see chapter 5).

**CHOOSING WEB CONTENTS AND FUNCTIONS**

The contents and functions needed for this project are the following:

- Basemaps

  - A street map: for basic spatial reference—from map services from ArcGIS Online

  - Aerial imagery: for better understanding of the ground environment—from map services from ArcGIS Online

  - Important resources and critical infrastructure: for locations of fire stations, police stations, airports, malls, and schools—from map services published via ArcGIS Server

  - Weather: for helicopter flight path—based on a map service from NOAA

- Operational layers

  - Highway incidents: from California Highway Patrol (CHP) Web reports

  - Highway traffic speed and camera: from California Department of Transportation (Caltrans), in text format

  - Helicopters (air ambulances and rescue airships): from a SOAP-based Web service by Air-Trak (Air-Trak 2006)

  - Ground vehicles (ambulances, fire trucks, etc.): available in SOAP-based Web services from Air-Trak (Air-Trak 2006)

  - Hospital diversion status: depicts the status and capacity of hospital emergency medical departments, available in XML format from the Hospital Association of Southern California (HASC) Rapid Emergency Digital Data Information Network Division (ReddiNet) Web service

- Tools

  - Find Address: from ArcGIS Online

  - Find Incidents: from a customized REST Web service

  - Other: while an actual project has its scope of work, other emergency response-related functions, such as routing based on the current highway speed, driving distance, and service area from a specific hospital, a plume model of hazard leaks, roadblock analysis, spatial statistical analysis against a series of disease locations, and so forth, can be integrated as needed

**DESIGNING SYSTEM ARCHITECTURE**

Not all of the preceding Web contents and functions are readily consumable by browser-side APIs. The highway incidents, traffic information, and locations of air ambulance helicopters do not have formal Web APIs; the ground vehicle and helicopter Web services are based on SOAP, which is not easy for browser-side programming languages to consume; and the hospital diversion status information is not georeferenced.

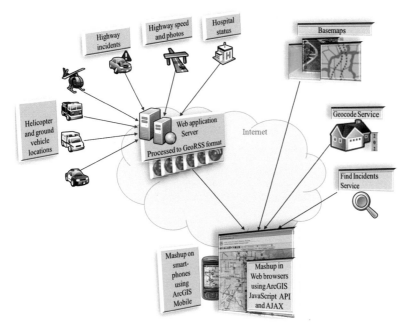

**Figure 4.14** This simplified version of the system architecture of the AEGIS application used by Loma Linda University Medical Center shows the use of a mashup for emergency response.

Courtesy of Loma Linda University Medical Center and Tele Atlas North America, Inc.

These contents need to be preprocessed through Web scraping or coordinate projection to yield easily consumable formats, such as GeoRSS. While mashups prefer easy-to-use browser-side programming (see section 4.2), the preprocessing end is usually complex and is done on the server side. The system architecture for this project uses .NET for server-side preprocessing and the ArcGIS API for JavaScript for browser-side mashup (figure 4.14).

## MASHING UP THE PARTS INTUITIVELY

A successful mashup needs to bring the Web resources together, present the results in an intuitive way, and make the application useful and usable. This is especially important when it comes to emergency response, when every second counts and the end users are the managers and executives who are making the potentially lifesaving decisions.

With the ArcGIS API for JavaScript, the basemaps and tasks served by ArcGIS Server are easily integrated in a mashup application. The operational data is preprocessed into GeoRSS format on the server side, and then pulled into the browser side dynamically using AJAX at regular intervals ranging from seconds or minutes for different feeds. The GeoRSS XMLs are further parsed to extract feature geometries and attributes used in the operational data layers.

The operational data layers in a mashup (figure 4.15) are usually displayed as graphics, and their symbols are dynamically determined by their attributes. AEGIS presents them intuitively—for

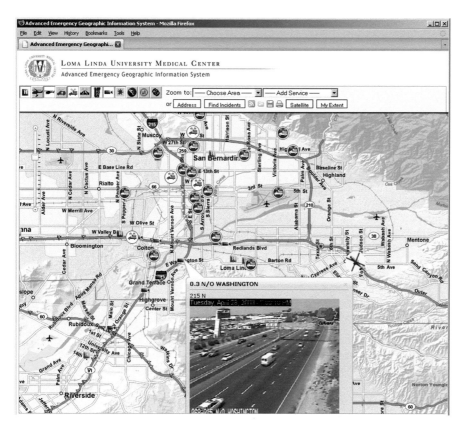

**Figure 4.15** Loma Linda University Medical Center's Web-based AEGIS combines continuous data feeds and photos in close to real time. Emergency managers can obtain comprehensive views of a situation, and query and zoom for more detail, to make crucial decisions quickly.

Courtesy of Loma Linda University Medical Center and Tele Atlas North America, Inc.

example, if a helicopter is in flight, an animation icon with rotating helicopter blades is used, and the direction of flight is indicated by the direction of the icon. If the emergency service at a hospital is at capacity, the hospital will be shown in yellow as a sign of caution to indicate that patients should be diverted to other hospitals. Vehicle speed detector sensors are displayed as green, yellow, and red flags, depending on average vehicle speeds. These graphics respond to mouse events. By hovering the pointer over a feature, a user can obtain additional detailed information, including near-real-time highway photos, in an information window.

## REAPING THE BENEFITS OF MASHUPS

The value of AEGIS lies in the comprehensive and real-time nature of information provided through a mashup. Information used in this system can be viewed at each individual source. For example, you can see the highway traffic information at the Caltrans Web site and the helicopter locations at the airport command center. Yet when all the information is integrated in one map, the application creates new value that cannot be achieved by looking at each source individually. The new overlaid map can reveal the spatial relationships among different incidents and different resources. LLUMC's emergency department can achieve situational awareness from one view integrating multiple sources. Emergency managers can make sound decisions quickly by addressing questions such as the following at a glance:

- Where are the patients? (Patients are tracked by field personnel using mobile devices, and located using the address finder function, or colocated with the traffic incidents.)

- Where are the air ambulances, rescue helicopters, and other emergency vehicles? Which one is closest to the patient?

- Which hospital is close to the patient with an emergency room that can currently accept this patient? (A hospital with a green "H" indicates that it accepts patients, while a yellow "H" means its emergency room is busy and the patient needs to be transported to another hospital.)

- What is the current highway speed in the vicinity? Is there any traffic congestion?

- Should this patient be transported by a ground vehicle or a helicopter?

AEGIS essentially achieved an overlay of disparate Web resources. Overlay is often challenged by the prerequisite that all input layers be compatible with each other and with the overlay software. It can take lengthy struggles with formats, projections, and spatial resolution before layers can actually be overlaid (Goodchild et al. 2010). These challenges existed for AEGIS as well. Highway incidents and traffic information had to be extracted from unstructured data sources using the Web scraping technique before they could be converted into standard GeoRSS feeds. The projection of some data sources was unknown and without metadata, so it took many phone calls and much experimentation to figure out the projection before the data could be projected into the coordinate system that matched the other layers. The hospital status feed had no spatial locations and so it had to be joined with a list of geocoded coordinates to be georeferenced. AEGIS extends from traditional overlays, however, by effectively dealing with data that is dynamic, distributed, and heterogeneous. It can do this through a mashup that takes advantage of the reach of the Web, the flexibility of Web services, standards such as GeoRSS, and the convenience of browser-side APIs.

## 4.4 CHALLENGES AND PROSPECTS

Mashups provide many opportunities for sharing data and using information in new applications, and the mashup has become a prevalent design pattern for Web GIS applications. You can ensure that your applications are useful, stable, and scalable with the use of best practices. As more and more contents and functions become available over the Web, and more techniques and policies favor the use of mashups, the potential of this methodology to create new value-added applications will grow exponentially in the future.

### 4.4.1 QUALITY, COPYRIGHT, CONSERVATION, AND SECURITY

Some important considerations need to be made in designing mashup applications. The following issues should be addressed:

- **Quality and authority of your sources:** Anyone can publish contents over the Internet, since there is no gatekeeper, so to speak. Quality can vary enormously, and if questionable sources are implicated in a chain of services, the results of uncertainty or error can be propagated over the Internet. This can produce misleading and even wrong information (Goodchild et al. 2010). The Web resources chosen should be appropriate for the purpose of the mashup application. Government or corporate mashup applications should use authoritative quality sources and pay attention to how the sources are processed and updated (see additional discussions about data quality and semantic interoperability in section 7.4.1).

- **Copyright and terms of use:** While the spirit of open exchange is still the main characteristic of the Web, many Web sites post terms of use and retain their copyright. Terms of use can include the possibility of including advertising information and the rate limit (i.e., the maximum number of queries and amount of access allowed). Copyright and terms of use can get complex when a mashup uses multiple Web resources, or even more complex when a mashup uses information generated by another mashup. It is important for mashup developers to comply with providers' terms of use to avert future disputes and to understand the impacts of copyright on their application.

- **Acting conservatively in the mashup ecosystem:** The consumers and providers of mashups form a mashup ecosystem, in which the actions of one affect the other. The stability of a mashup application depends on the stability of its sources. Conversely, a mashup puts new loads on its sources, which will affect source stability. For example, if Web service A is consumed by 10 mashup applications, each mashup application gets 100 clicks per minute. All the hits (100 x 10 = 1,000 per minute) may be loaded into service A, which may exceed the planned capacity of service A and thus bring it down. This in turn can bring down all the mashup applications that depend on service A.

  Providers need to plan ahead for the additional traffic that mashups produce, and the mashup developer needs to act conservatively to reduce the pressure on the provider. Mashup developers can use the server caching technique—that is, have a Web server read the source once and keep it for a certain period of time for all browsers to use, rather than letting each browser call the Web source separately.

- **Security:** Enterprise mashups can involve confidential information, making security a key consideration. The need for security has been briefly discussed in the comparison of server-side versus browser-side mashups earlier in this chapter. Security for mashup applications can be enforced on the service provider side (for Web services security, see section 3.5.5) or the mashup application side—for example, requires login, doesn't leave the browser-side cache, and uses HTTPS.

## 4.4.2 THE POTENTIAL FOR EXPONENTIAL GROWTH

The orchestration of Web services into a workable composition has been an important research topic since Web services originated in the late 1990s. The emergence of browser-side APIs in 2005 has been the driving force for the success of mashups. Mashup Web sites have changed both the developer's and the Web user's experience. For developers, browser-side APIs have greatly reduced the effort needed to combine multiple Web services. For Web users, browser-side APIs provide rich and responsive interaction that has changed Web users' expectations of Web GIS. These features have resulted in an explosion of public interest in using the Web to spontaneously create, assemble, and disseminate geographic information. The design pattern used in mashups has proven effective for both casual applications and critical enterprise Web projects. GIS software vendors provide extended functionality to support the prevailing pattern of mashup design.

Mashups are associated with many areas of geospatial science and technology, including information sharing, geospatial information mining, geospatial information interoperability, and the next generation of National Spatial Data Infrastructure (see chapter 7). As an easy way to use the Web resources available on other people's computers, including their hardware, software, data, modeling, and knowledge, mashups also have given impetus to advances in such frontiers as cloud computing (see chapters 7 and 10).

While the mashup process has been widely recognized as easy, this is primarily true of mashups using Web resources with formal APIs. The billions of Web pages in existence contain a huge amount of data that can be used in boundless practical applications—that is, if the data can be extracted and formatted. These resources are still largely untapped because of the difficulties in Web scraping and geoparsing. Web resources are unevenly distributed, and most of the high-resolution basemaps available over the Web are for North America and Europe, with only limited areas (i.e., major cities) of the rest of the world. For some regions of the world, it is still difficult to find a decent basemap to use in a mashup.

In the future, developers will be able to make more efficient use of Web resources when they can understand the essence of Web contents and functions more accurately and more easily discover the Web resources that best fit their applications. This will become possible through the improvement of geospatial Web portals (see chapter 6) and the realization of the Semantic Web (see chapter 10). With the development of the next generation of NSDI, governments, companies, and institutes are increasingly sharing information via Web services, which is a boon for society. Sensor networks in space, on the ground, and under the ocean are collecting huge volumes of geospatial data, and this observation data will be widely accessible via standard Web protocols. Citizens in general are voluntarily reporting what they see and what they feel with the aid of their cell phones, social networks, and other Web applications. As the collection of georeferenced contents multiplies, the ways to combine such information will grow exponentially. The Web user community and GIS professionals will create more innovative and value-added Web GIS applications to serve a broad array of societal needs.

## Study questions

1. What are mashups and geospatial mashups?

2. When did mashups emerge, and how did the technology evolve?

3. A typical geospatial mashup has basemaps, operational layers, and tools. What is the role of each type of component?

4. What is a Web API? What different types are there? Compare the benefits of each.

5. Visit the ESRI Resource Center Web site. Try out the sample applications of the ArcGIS APIs for JavaScript, Flex, and Silverlight. Explain the advantages of browser-side APIs.

6. List some commonly used Web contents and functions for geospatial mashups.

7. What is Web scraping? What is it used for?

8. Imagine that you need to develop an Earthquake Disaster Alert application. In this application, you will need to display near-real-time earthquakes on basemaps. For earthquakes with a magnitude greater than 6, you will need to report the cities, counties, and rough number of population that will be affected. You will need to address the following issues:

   • What contents and functions are needed?

   • Where would you find them?

   • Which programming interfaces do they use?

   • What are your basemaps, operational layers, and tools?

9. The mashup method suits almost all areas of Web GIS applications. Why is it especially fitting for situational awareness and executive dashboard types of applications?

10. What impacts do mashups have on GIS?

## References

Air-Trak. 2006. The Advanced Emergency Geographic Information System (AEGIS). `http://www.air-trak.com/case-studies-llumc.php` (accessed February 19, 2010).

Brown, Clint. 2009. Building great Web maps. ArcGIS Developer Summit, Palm Springs, Calif., March 23–26.

Brown, Clint, Jeremy Bartley, Ismael Chivite, Bernie Szukalski, and Rob Shanks. 2008. ArcGIS in a Web 2.0 World. ESRI International User Conference, San Diego, Calif., August 4–8.

Fike, Ruthita. 2007. How GIS is changing Loma Linda University Medical Center's view of the world. *ArcNews* (Fall). `http://www.esri.com/news/arcnews/fall07articles/loma-linda.html` (accessed November 15, 2009).

Goodchild, Michael F. 2007. Citizens as sensors: The world of volunteered geography. *GeoJournal* 69 (4): 211–21.

Goodchild, Michael F., Pinde Fu, Chaowei Yang, and Jeff Grange. 2010. Geospatial mashup: Technologies, challenges, and opportunities. (Manuscript).

Hanna, Andy. 2008. Loma Linda University Medical Center's Geographic Information System. National Academy of Public Administration. `http://www.collaborationproject.org/pages/diffpages.action?originalId=6389926&pageId=6389927` (accessed February 19, 2010).

Kolbasuk McGee, Marianne. 2008. Hospital implements high-tech emergency response system. *InformationWeek,* June 25. `http://www.informationweek.com/news/personal_tech/gps/showArticle.jhtml?articleID=208800864` (accessed February 19, 2010).

Shah, Nirav B., Donna H. Rhodes, and Daniel E. Hastings. 2007. Systems of systems and emergent system context. Proceedings of CSER, Hoboken, N.J., March 14–16. `http://web.mit.edu/adamross/www/SHAH_CSER07.pdf` (accessed February 19, 2010).

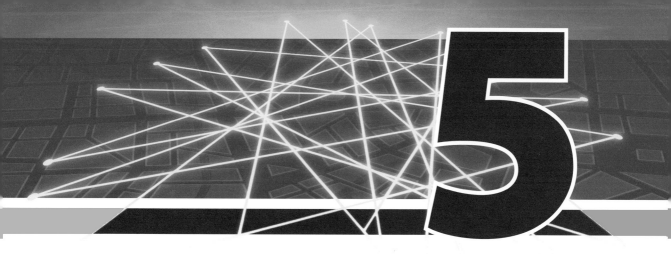

# MOBILE GIS

Wireless communications and mobile computing look to be the key technologies of the twenty-first century. With more than 4 billion mobile cellular subscribers (ITU 2009), wireless communications and mobile computing have gained acceptance worldwide with a speed that has surpassed many other technical innovations. The proliferation of cell phones and other mobile devices, along with the fundamental need for convenience, have brought about the rapid progression of mobile GIS—that is, GIS for use on mobile devices.

It is important to introduce mobile GIS in the context of Web GIS. Mobile GIS emerged in the mid-1990s to meet the needs of field work such as surveying and utility maintenance. Such early systems operated mainly in a disconnected mode. But with radical advances in wireless communications, especially the operation of 3G networks, mobile GIS is increasingly connected to the Web, and thus becoming a part of Web GIS. Mobile GIS can update the server with the latest information from the field. The Web server can, in turn, support mobile GIS with rich content and advanced analytics.

Mobile GIS is needed by the general public just as it is by organizations and professionals. Consumer mapping applications used on the mobile platform provide location-based services (LBS) related to the location of the user. This gives you the convenience of using GIS on the go—to find things that are nearby, to get to places, and to locate friends. Mobile GIS can potentially be used by anyone, for anything, anytime, anywhere.

This chapter introduces the concept of mobile GIS, along with its uses and benefits. Section 5.2 presents the supporting technologies of mobile GIS, including the mobile platform, wireless communications, and positioning technology. Section 5.2 also discusses the technical challenges facing mobile GIS. Section 5.3 summarizes mobile GIS applications development options, including native-application-based, mobile-browser-based, and message-based solutions. Section 5.4 presents a number of case studies in field survey, data collecting, asset inspection, emergency response, and LBS. Section 5.5 discusses the social, industrial, and technical issues regarding mobile GIS along with its prospects.

## 5.1 USES AND BENEFITS

In the past decade, there has been no greater influence on GIS architecture than in the enormous improvements that have been made in computer networks (Maguire 2007). We are in the midst of a radical shift from wired to wireless communications, and from wired to wireless computer networks. Without the constraint of wires, GIS can be used in the field and wherever else it's needed. This flexibility holds many advantages for industry workers and for consumers.

### 5.1.1 FROM WIRED TO WIRELESS

Geospatial information is needed in virtually all walks of life, although many people don't realize it. GIS is needed on mobile or wireless devices used on the street and in the field just as it is needed on desktop computers wired into the home and office. You need GIS in your daily life for such things as finding the best route to places you want to go, finding places to eat when you are hungry, and finding out where to go when you are lost.

Many business tasks can't be completed from the office. Such tasks as field surveys, product deliveries, power line maintenance, utility troubleshooting, and emergency response require the crew to step out of the office and into the field. To accomplish these tasks, people in the field need to have maps in hand to facilitate asset inventory and inspection, to find the location of gas or water pipes so they don't dig up the wrong ones, to provide driving directions so they can optimize a route of multiple stops, and to familiarize themselves with an incident area so they can respond effectively (figure 5.1). All these needs have culminated in a new field of GIS—mobile GIS.

**Mobile GIS refers to GIS for use on mobile devices.**

One unique aspect of mobile GIS is that users in the field are usually in the same area as the subject (Longley et al. 2005). Conceptually, there are four locations in a GIS: the location of the user (U), the location where the data is stored (D), the location where the data is being processed (P), and the area that is the subject of the GIS project (S). In traditional GIS, the user usually sits in the office, and the subject can be anywhere in the world—thus, U ≠ S. **In mobile GIS, U = S. That is, the user is usually located in the subject area; thus, the user can see, touch, feel, hear, and even smell the subject.** For example, first responders are right at the incident scene, and they can record what they see, feel, hear, and smell. Such information extends the awareness of those in command through observations from afar.

**A**

**B**

**C**

**D**

**Figure 5.1** GIS is needed on the road and in the field for such functions as (A) orientation in a new city, (B) asset inventory and inspection, (C) in-vehicle navigation, and (D) emergency response.

The early applications of mobile GIS in the 1990s were often used for in-vehicle navigation and field surveying. GIS data and software were preloaded and operated in a disconnect mode in the field. With the advances in wireless communications over the last decade, more and more mobile GIS applications are connecting to the Internet. These applications use the rich and powerful GIS services available over the Web, gain updates of the latest GIS data, and regularly post the latest field information back to the office. **Mobile GIS has evolved from being disconnected to wireless connected and is increasingly becoming a component of Web GIS.** The study of mobile GIS overlaps the topic of Web GIS (figure 5.2), and it is the reason why mobile GIS is included in this book.

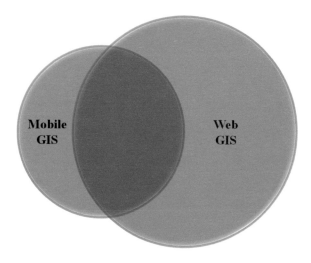

**Mobile GIS**

**Web GIS**

**Figure 5.2** Early mobile GIS operated mostly as a stand-alone system. With the adoption of wireless communications, mobile GIS is increasingly being connected to the Web, and thus overlaps the topic of Web GIS.

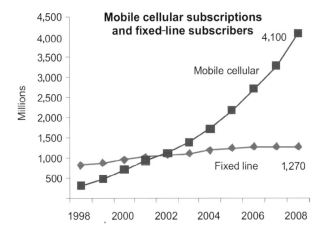

**Figure 5.3**   The number of mobile phone subscribers worldwide has grown rapidly over the past ten years.

Courtesy of International Telecommunication Union.

The popularity of mobile devices and the increased speed of data transfer using wireless communications have led to expanded adoption of mobile GIS. Many organizations equip their staffs with mobile devices to aid the workflow, and many individuals buy mobile devices to satisfy their need for convenience. The number of subscribers of mobile communications has grown at a rate of about 20 percent a year this century. According to the International Telecommunication Union, the number of mobile phone subscribers surpassed fixed-line subscribers in 2000 (figure 5.3) and was set to reach 4.1 billion, more than half the world's population, by the end of 2008 (ITU 2009) and to reach 5 billion in 2010 (ITU 2010). This huge volume of mobile phones and other mobile devices can potentially be used as a platform for mobile GIS through text messaging, Web browsing, and native applications.

## 5.1.2 FUNCTIONS AND APPLICATIONS

Mobile GIS is well suited to some GIS functions, but not so well suited for others. Mobile GIS works well with the following GIS functions:

- **Information capture and updating:** Mobile GIS is an ideal platform for capturing information and updating it. Mobile devices are portable, and many have built-in GPS receivers, which allow them to acquire accurate location information in addition to attribute information. Contemporary mobile devices come with cameras, voice recorders, and even video cameras, enabling them to collect rich geospatial information in several multimedia forms.

- **Dissemination:** Mobile devices can bring GIS into the hands of billions, whether users are in the field, in the office, in the home, or on the go.  You can view maps and query what is around you on demand.

Mobile GIS is relatively weak in the following GIS functions:

- **Storage:** Except for especially rugged units, most mobile devices are not considered a robust platform for large databases. Data is usually stored temporarily on a mobile device and needs to be synchronized with a GIS server after a day's work, or on a more frequent basis via wireless communications.

- **Analysis:** Because of the limited computing power of a mobile device, it cannot provide the comprehensive and sophisticated analysis capabilities of a desktop computer or server. Mobile GIS applications often have specialized functions for specific purposes, such as routing for in-vehicle navigation. Other advanced analyses are usually performed remotely on a Web GIS server rather than being done locally.

- **Presentation:** Mobile screens cannot compare with desktop monitors in terms of size of presentation and depth of color. The user interface and map presentations of mobile GIS are usually simplified for easy operations and readability.

## CONSUMER MOBILE GIS APPLICATIONS

Consumer mobile GIS applications generally answer questions related to daily life such as "Where am I?", "What is around me?", "Where is the closest restaurant, bank, ATM, gas station, or store?", and "How do I get from here to there?" These applications usually fall under the umbrella of LBS (see section 5.4.3). **LBS refers to information services that integrate the location of a mobile device to provide added value to the user** (Spiekermann and Humboldt-Universitat 2004).

Consumer applications share certain similarities in that they all need to have basemaps and points of interest (POIs), and they all need to have map, query, and routing functions. The main competition among service providers focuses on who has the most comprehensive and up-to-date basemaps and POIs, and who has the best usability and trendiest user interface. These consumer applications are usually provided free by advertising companies or other organizations that charge for services through subscription fees. Typical examples of consumer applications include Google Maps for Mobile, Microsoft Bing for Mobile, Yahoo for Mobile, and MapQuest Mobile.

## ENTERPRISE MOBILE GIS APPLICATIONS

Many organizations, including utility and infrastructure companies, public safety and law enforcement agencies, and government organizations have large teams composed of utility workers, surveyors, engineers, police officers, firefighters, soldiers, census workers, and field biologists. These field crews need mobile GIS (ESRI 2007a) to complete tasks such as the following:

- **Field mapping, query, and decision support:** View maps and query geospatial features in the field to provide information for decision making.

- **Field inspection and inventory of assets:** Inspect assets such as transformers, water meters, street signs, and home improvement projects. Mobile maps and GPS help field crews ensure that locations and information are accurate. The digital records collected can be maintained for legal code compliance and ticketing.

- **Field surveying:** Use the GPS of mobile devices to achieve high horizontal and vertical resolution and reduce the time lag between data collecting and dissemination.

- **Incident reporting:** Document and report the location and circumstances of incidents or events.

- **Collaboration:** Coordinate communication between workers in the field and in the office.

- **Tracking:** Track and report the location of mobile devices to other mobile devices or to the command center. A mobile user can locate another mobile user, and the command center can track all mobile users, vehicles, and other mobile users.

Enterprise applications have a greater scope of functionality than consumer applications. Each application has a specific purpose and requires specific data. For example, a pipeline field worker needs GIS data about the pipeline network and maps of local streets to be able to trace a path to the right valves, turn them off, and then dig at the correct spot. After a disaster, employees of an insurance company need to have a list of customers and their locations, so they can drive to the correct spot and collect information about property damage to process insurance claims. A first responder needs to have a relevant basemap, information on critical infrastructure, and various live feeds to gain situational awareness and may even need the floor plans of a building to perform certain rescue tasks.

In an enterprise application, an organization generally procures mobile devices for its staff, collects GIS data, acquires mobile GIS software, and customizes or configures the software to meet specific business requirements. The main considerations are whether the application fits the specific needs of the workflow, whether it is easy for employees to use, whether it can be connected to the office system to post data and synchronize tasks, e-mails, and calendars, and whether it is secure. Whether it is hip and trendy is not usually among the most important considerations of an enterprise application.

### 5.1.3 ADVANTAGES OF MOBILE GIS

Mobile GIS has certain advantages over traditional desktop GIS. Among them are the following:

- **Replacing paper-based workflows:** Mobile GIS can replace existing paper-based workflows, which can be prone to error from people manually entering and interpreting written information and can involve redundant work processes. Paper-based workflows can also inhibit the currency of data. By leveraging the power of a mobile device, you can replace paper forms, reduce costs, and improve the currency and accuracy of your data.

- **Mobility:** Mobile devices are not hindered by wire or cable and are usually compact and portable. Most of them can be held in the hand or the pocket. Larger units can be stuffed in a backpack or mounted in a vehicle. Specially manufactured mobile devices can work in harsh or hazardous environments. Mobile devices can extend GIS to areas where wiring is infeasible or costly, where it is too hot or humid for normal computers to function, and to places where users need GIS the most.

- **Large volume of users:** With more than 4 billion cell phone subscribers, plus a broad usage of other types of mobile devices, mobile users have far exceeded the number of desktop computer users. This large volume of users is making mobile devices a pervasive, and even ubiquitous, platform for GIS. The popularity of mobile devices provides the opportunity for GIS to be accessible to the largest population possible. It broadens the market for the geospatial industry and provides tremendous potential for the use of geospatial science.

- **Location awareness:** A number of technologies can be used to pinpoint the current location of a mobile device (see section 5.2.3). The ability to know "where I am" is a unique characteristic of mobile GIS. It is what allows mobile GIS to be used in survey, navigation, emergency rescue, finding POIs, and other LBS.

- **Versatile means of communication:** Users of mobile GIS can communicate via voice, short message, photo, video, e-mail, and the Web. This versatility facilitates collaboration between field workers and the office, improving the accessibility of information and increasing the efficiency of the workflow.

- **Near-real-time information:** The application domain of mobile GIS is usually wherever an event is taking place and whenever a user needs information. Field events can be recorded and reported to the Web server in near real time. This capability greatly enhances the temporal dimension of GIS. Mobile GIS has the potential to monitor the spatial and temporal aspects of the world around us.

## 5.2 SUPPORTING TECHNOLOGIES

Mobile computing, wireless communications, and mobile positioning are among the most exciting advances in information technology. These technologies (figure 5.4) form the foundation of mobile GIS. This section introduces these technologies as well as the technical challenges they pose for mobile GIS.

**Mobile phones**

**Pocket PCs**

**Portable PCs**

**Special devices**

**Figure 5.4** Mobile devices come in four categories.

Blackberry, RIM, Research in Motion, SureType, SurePress, and related trademarks, names, and logos are the property of Research in Motion Limited and are registered and/or used in the U.S. and countries around the world. Courtesy of Trimble.

### 5.2.1 MOBILE DEVICES

Mobile devices can be classified in four categories: mobile phones, pocket PCs, portable PCs, and other special devices such as in-vehicle GPS navigation systems.

#### MOBILE PHONES

Mobile phones include cell phones (connecting to terrestrial cellular sites) and satellite phones (connecting to orbiting satellites). Mobile phones are the most common mobile devices. Low-end mobile phones usually have very limited local storage and processing capabilities. Smartphones offer advanced capabilities, often with PC-like functionality. In addition to the standard voice function, many smartphones support Short Message Service (SMS), Multimedia Messaging Service (MMS), Web browsing, e-mail access, and instant messaging. Mobile phones increasingly have built-in Bluetooth capabilities, a camera, and GPS. Low-end mobile phones can access geospatial information via messaging services. Some smartphones can browse Web GIS sites, install GIS applications, and load data.

#### POCKET PCs (POCKET COMPUTERS, PALM COMPUTERS)

Pocket PCs, also called PDAs (personal digital assistants), are handheld computers. Despite their small size, pocket PCs provide a strong platform for mobile GIS.

#### PORTABLE PCs (TABLET PCs, LAPTOPS)

Portable PCs are actually a full-featured computer in a portable size. Portable PCs have advanced processors and large data storage. They are capable of running full-featured desktop GIS products. Portable PCs are relatively heavy compared with other mobile devices and are often used in a vehicle.

#### SPECIAL DEVICES

Special devices include a variety of embedded devices, including in-vehicle GPS navigation systems.

Similar to desktop computers, mobile devices have a mobile operating system (OS) that determines what functions and features are available, controls the thumbwheel and the keyboard, and handles wireless connections. Common mobile operating system platforms include Symbian, iOS, Windows Phone, Google Android, and Blackberry. This large variety of platforms introduces a challenge to the development of mobile GIS in that a mobile GIS application may not be able to run consistently on all platforms. Mobile GIS application development needs to consider which mobile OS platforms the targeted audience uses, and then design the application for the selected platforms.

**Figure 5.5** Wireless communications technologies vary in their range.

## 5.2.2 WIRELESS COMMUNICATIONS

Wireless communications technologies vary in their speed, range, and setup costs, and fulfill different roles (figure 5.5). For the ease of discussion, this section refers to the typical specifications for Bluetooth, Wi-Fi, and cellular networks.

### BLUETOOTH

Bluetooth, named after Danish king Harold Bluetooth, is designed for short-range communications of about 10 meters. Its applications include communication between a mobile device and its peripherals such as a GPS receiver and headset.

### WI-FI (WIRELESS FIDELITY)

With a typical range of 100 meters and a data transfer rate of 10 to 54 MBps, Wi-Fi is commonly used to create a wireless local area network (WLAN) within a family, a company, a university, or public area such as a library, airport, railway station, or coffee shop. Areas that have Wi-Fi connections available are commonly referred to as "hot spots."

The main disadvantage of Wi-Fi is its limited spatial coverage. Wi-Fi hot spots are limited to 100 meters from the wireless router, and the coverage of Wi-Fi hot spots is not contiguous.

### CELLULAR NETWORKS

A cellular network is a type of radio network made up of a number of radio cells, each served by at least one fixed-location transceiver known as a cell site or base station. These cells cover different land areas to provide phone coverage over a wide area. There are different cellular networks set up by wireless carriers and government agencies. The size of a cell is based on such factors as radio frequency, terrain, and the number and proximity of potential customers. Cellular network technology has evolved across the following generations:

- 1G: First-generation wireless telecommunications operates on analog technique and transmits voice only. Mobile GIS cannot be built on this technology.

- 2G and 2.5G: Second-generation technology operates on digital cellular networks. It supports voice services and low-speed data transmission services under 10 KBps. 2.5G is a stepping-stone between 2G and 3G. 2.5G updates the current 2G network and can achieve faster data transmission speeds up to 384 KBps. Building mobile GIS on 2G and 2.5G networks faces a severe challenge of limited data transfer speeds.

- 3G: Third-generation technology is currently available in many countries. It offers data transmission speeds up to 2 MBps and supports a wide range of advanced services, including Web surfing and video streaming. With 3G technology, mobile clients can surf GIS Web sites, integrate geospatial Web services, and provide a smooth user experience.

- 4G: Fourth-generation technology—that is, the next generation—will provide data transmission speeds of about 100 MBps. The backbone of the network will probably be built on LTE (Long-Term Evolution), WiMax (Worldwide Interoperability for Microwave Access), or UMB (Ultra Mobile Broadband) technologies. 4G will have a data speed that is close to what is being offered by the current LAN, so it will allow deployment of mobile GIS capabilities as efficient as what is being used in the office today. Mobile GIS will thrive in the 4G era.

### 5.2.3 MOBILE POSITIONING TECHNOLOGY

**The cornerstone of most mobile GIS applications is location awareness. This is particularly important for LBS such as emergency rescue.** Knowing the location of a mobile device can help pinpoint the location of its owner or of an incident involving the owner. That information was critically lacking in the case of a woman who lost control of her car and skidded into a canal in Florida in 2001. She used her cell phone to call for emergency help but couldn't describe her exact location, and the operator was unable to locate her address by cell phone. The rescue units did not know where to go to find her, and the woman died before her car was found (*Popular Science* 2002). Realizing that the large and growing percentage of emergency calls are made by cell phone, countries have passed legislation such as Enhanced 911 (E911) in the United States in 1996 and similar legislation in Europe, which require mobile phone providers to provide the location of emergency calls with a certain degree of accuracy. These requirements have driven considerable advances in mobile positioning technology.

**The main mobile positioning technologies today are based on satellite, cellular network, Wi-Fi network, IP address, and radio frequency identification (RFID) locations.**

#### NAVIGATION SATELLITE-BASED APPROACH

GNSS (Global Navigation Satellite System) allows small electronic receivers to determine their location (longitude, latitude, and altitude) by using the signals transmitted by satellites. The most commonly used satellite system is the U.S. Global Positioning System (GPS). Other satellite systems include Russia's GLONASS, China's Beidou system, and the European Union's Galileo system, which is in its initial phase of deployment.

The satellite approach is also called the handheld approach because it requires an internal or external satellite receiver on the mobile device as well as a native program to access the receiver. The advantage of this approach is accuracy of location. The location acquired from GPS under a clear sky has an average horizontal accuracy of 5 meters (Kowoma 2009). The horizontal accuracy can reach submeter accuracy with real-time differential correction and subfoot accuracy with postdifferential correction depending on the type of GPS receiver. This approach is often used for applications where high accuracy is required, such as in-vehicle navigation, field survey, and utility maintenance. The disadvantage is that satellite-based accuracy can be reduced by satellite position, cloud cover, and physical barriers such as high-rise buildings.

### CELLULAR-NETWORK-BASED APPROACH

The cellular-network-based positioning approach relies on the way the cellular site finds mobile devices in its service territory and routes calls or other types of communication services to them. This approach has relatively low accuracy but can still support a variety of consumer LBS applications.

The network-based approach can be implemented via several different methods. The most basic method is cell of origin (COO), which considers the location of the origin base station as the location of a caller. This method requires little investment, but its accuracy depends on the size of the cellular cell, which varies from a few hundred meters in urban areas to tens of kilometers in rural areas. Other methods include signal time of arrival (TOA), time difference of arrival (TDOA), angle of arrival (AOA), and enhanced observed time difference (E-OTD). These methods rely on triangulation of the cellular network and can reach location accuracies within 100 meters, but they need expensive equipment such as directional antennas and accurate clocks in the cellular network system.

### ASSISTED GPS APPROACH

The Assisted GPS (A-GPS) approach uses satellite-based and cellular-network-based technology. It can achieve accurate location and still operate in areas behind or under barriers.

### WI-FI-BASED APPROACH

When a mobile device connects to the Internet via Wi-Fi, its location can be determined by the location of the Wi-Fi hot spot. This method relies on a database of Wi-Fi hot spots and the fact that a mobile device must be within 100 meters of its Wi-Fi access point. With Wi-Fi signal triangulation technology, the accuracy can be higher for certain areas. The limitations of this approach are that it does not work when the mobile device is out of range of Wi-Fi signals and that the Wi-Fi hot spots database must constantly be updated.

### IP ADDRESS-BASED APPROACH

When a mobile device connects to the Internet, the IP address of the mobile device, which is usually a gateway of the Internet service provider (ISP), can be used to determine the location of the mobile user. The gateway is an area that is shared by many households or organizations. The accuracy of this approach is relatively low, usually within the range of a city area.

## 5.2.4 TECHNICAL CHALLENGES

Mobility comes at a price. The small size of mobile devices imposes limitations on CPU speed, memory size, network connections, screen resolution, and power supply. Some of these limitations should become less severe as the technology advances, but they still must be taken into account when designing mobile applications. Mobile GIS application design and development should be aware of the following challenges:

## LIMITED SYSTEM RESOURCES (CPU, MEMORY, AND BATTERY POWER)

Because it runs on limited system resources, mobile GIS should have minimal installation and be as efficient as possible. Complex processing should be left on the server side as Web services for the mobile application to use on demand. Data that has to be preloaded should be kept to a minimum, with unnecessary attribute fields removed and the geometry simplified. Data collected by mobile devices needs to be synchronized to the server at regular intervals to prevent unexpected power failure and data loss.

## LIMITED BANDWIDTH AND INTERMITTENT NETWORK CONNECTIONS

Physical conditions, such as weather and the surrounding environment, can affect the quality of service of wireless connectivity. The actual speed of a wireless network is usually slower than its theoretical maximum. Some areas may not be covered by the wireless communication network. Users may experience insufficient bandwidth or intermittent disconnection. For applications that need to operate when disconnected, the mobile GIS program and data should be preloaded locally. For applications that rely on network connections, the data to be transferred across the mobile network should be kept to a minimum size.

## SMALL SCREEN SIZE, TINY KEYBOARD, AND OUTDOOR ENVIRONMENT

Extenuating factors challenge the user interface design of mobile GIS applications. For example, 320 by 240 pixels is the most common screen resolution for mobile devices, but that is less than 10 percent of the normal display space of desktop computers. While some mobile devices have touch screens, the accuracy of a person's fingertips is usually low.

The user interface of mobile GIS should focus on field tasks and workflows. Designers should avoid replicating a desktop user interface on a mobile device and leverage the hardware capabilities of the device—for example, soft keys and a touch screen. Rather than cluttering a display with toolbars and menus, use forms or pages to display content. Decisions on data entry should be made where possible for the field user, and text entry dialogs should be avoided. Instead, leverage the power of the data model and present a list of choices where needed. Consider environmental conditions when choosing fonts and colors. The effect of direct sunlight on color choices (do not design for indoor use) and the darkness effect of nighttime use should be considered. Maps should use bold text labels, high contrast, and simple symbology so that they can be easily recognized. In most cases, leverage the location capabilities of the mobile device for both map navigation and search. For street navigation-type applications, the map should be oriented toward the direction of the user's movement rather than due north; the map designer should consider using 3D (or 2.5D) display instead of plain 2D; and driving directions should use voice commands instead of text, since the user will likely be driving.

In summary, the mobile platform presents many opportunities as well as challenges for GIS. The bottom line is, the mobile platform is important, evolving, and increasingly popular. Regardless of its limitations, mobile GIS is becoming an integral form of Web GIS, and it offers many advantages.

# 5.3 SOLUTIONS AND PRODUCTS

Mobile GIS can be developed using several approaches (table 5.1), including native applications, mobile browsers, and messaging (Kapoor 2008). This section discusses the pros and cons of each, and helps readers make the right choices based on an application's purpose, requirements, and target audience.

| | NATIVE APPLICATION BASED | | MOBILE BROWSER BASED | | MESSAGE BASED |
|---|---|---|---|---|---|
| | **WITH DATA LOADED** | **WITHOUT DATA** | **WITH BROWSER PLUG-IN** | **WITH WAP OR HTML** | |
| **APPROACH** | Both software and data reside on the device | Software only resides on the device | With a plug-in such as Flash Lite, Silverlight Mobile, or Java FX Mobile | View simplified or full-featured HTMLs in mobile browsers | SMS and MMS |
| **ACCESS TO GPS RECEIVER** | Yes | Yes | No | No | No |
| **DISCONNECTED OPERATION** | Yes | No | No | No | No |
| **DEVICE RANGE** | Limited | Limited | Limited | Broad | Very broad |
| **USER EXPERIENCE** | Good | Good | Good | Fair | Poor |
| **DEVELOPMENT COST** | High | High | Medium | Low | Low |

**Table 5.1**  Comparison of approaches to mobile GIS application development

## 5.3.1 NATIVE APPLICATION APPROACH

Native applications, in which software programs physically reside on and run on the mobile device, are the main approach for mobile GIS applications development. Native programs are often written in languages such as .NET Mobile or Java Mobile Edition and need to be downloaded and installed on the mobile device before they can be used. This approach requires middle to high-end mobile devices. While this may limit the audience, there are also many advantages:

- Ability to access mobile peripheral devices such as GPS receivers and to acquire accurate location information. This is especially important for applications such as surveying, navigation, and many enterprise mobile applications.

- Ability to access local files and databases. For applications that must operate when wireless communications are not available or cannot be guaranteed, the required data must be deployed on the mobile device. Native applications are needed to access local data.

Native applications are used in both consumer products and professional applications. Examples of consumer applications include in-vehicle navigation systems, Microsoft Bing Maps for Mobile, Google Maps for Mobile, and most applications available at the App Store of Apple Inc. and the Android Market of Google. These applications may work only on certain mobile operating systems depending on the programming languages they are developed in. Examples of professional products include ESRI's ArcPad, ArcGIS Mobile out-of-the-box applications, ArcGIS for iPhone, and ArcLogistics Navigator.

ArcGIS Mobile comes with an out-of-the-box application and a software development kit (SDK). The configurable out-of-the-box ArcGIS Mobile application (figure 5.6) integrates with ArcGIS Server to provide central management and deployment of mobile GIS data, maps, tasks, and projects. The ArcGIS Mobile out-of-the-box application enables users to do the following:

- View and navigate mobile maps to direct field resources effectively and monitor the location of assets

- Collect, edit, and update GIS data in real time and share information directly with colleagues

- Search and manage a list of GIS features to perform tasks or plan future work

- Rapidly deploy mobile GIS without having to develop a new solution

ArcPad is designed for professionals who require GIS capabilities in the field. It provides field-based personnel with the ability to capture, edit, analyze, and display geographic information efficiently (figure 5.7). ArcPad enables users to do the following:

- Perform reliable, accurate, and validated field data collecting

- Integrate GPS, range finders, and digital cameras into GIS data collecting

- Share enterprise data with field workers for updating and decision making

- Improve the productivity of GIS data collecting

- Improve the accuracy of the GIS database and perform updates

The ArcGIS for iPhone application and the ArcGIS Mobile SDK expand the reach of maps published with ArcGIS Server to iPhone (figure 5.8). The iPhone application can display maps from ArcGIS Online or from your own ArcGIS server, find work orders and customer calls, and act as a sensor that feeds an enterprise GIS with locations and observations.

ArcLogistics Navigator runs on Tablet PCs and Windows Phone and is a GPS-guided in-vehicle navigation product that helps mobile professionals stay en route and on schedule. ArcLogistics Navigator is designed specifically for fleets and mobile workforce applications and can be integrated with ESRI routing and mobile products to create scalable logistics solutions. It can help drivers minimize missed turns, reduce mileage, and shorten drive times by providing audible, turn-by-turn directions. ArcLogistics Navigator operates differently from consumer, off-the-shelf GPS products. A dispatcher can use it to send optimized routes and sequenced stops, with current data about each stop, to devices running ArcLogistics Navigator directly from ESRI routing products. The driver is guided to each stop along a preset optimized route, which honors all logistics-specific restrictions such as one-way streets, curb approach, height restrictions for bridges and tunnels, and weight restrictions for bridges.

The native application approach typically needs a programming language such as .NET Mobile or Java ME, as well as a GIS SDK such as ArcGIS Mobile SDK, ESRI iPhone SDK, or ArcGIS Engine. The ArcGIS Mobile SDK provides a set of .NET programming components for GIS data display, map navigation, GPS support, geometry and attributes editing, and synchronization with ArcGIS Server.

**Figure 5.6** ArcGIS Mobile helps organizations deliver GIS capabilities and data from centralized servers to a range of mobile devices for such functions as (A) finding coworkers or facilities, (B) performing tasks via a task-driven user interface, (C) editing attributes and appending photos, or (D) using the Tablet PC version of ArcGIS Mobile in "night skin."

Map data © AND Automotive Navigation Data; courtesy of Tele Atlas North America, Inc.; GeoEye; i-Cubed; and Trimble.

**Figure 5.7** GeoCollector, which is ArcPad bundled with the Trimble GeoExplorer 2008 Series professional GPS handheld device, can achieve subfoot accuracy in field surveying.

Courtesy of Tele Atlas North America, Inc., and Trimble.

**Figure 5.8** ArcGIS for iPhone can act as a Web client for ArcGIS Server, with the ability to perform searches (left and middle), display maps (right), collect field data, and request server models.

Map data © AND Automotive Navigation Data; courtesy of Tele Atlas North America, Inc.

## 5.3.2 MOBILE BROWSER APPROACH

A mobile browser, also called a microbrowser, minibrowser, or wireless Internet browser (WIB), is a Web browser designed for use on a mobile device. Mobile browsers are specially designed to effectively display Web content on small screens on portable devices. Mobile GIS can run in mobile browsers with HTML, a subset of HTML, or a mobile plug-in.

### WAP (WIRELESS APPLICATION PROTOCOL)

Realizing the limitations of the mobile platform, the international standards bodies devised WAP, a standard for application layer network communications in a wireless communications environment. Its main use is to enable access to the mobile Web from a mobile phone or pocket PC. WAP 1.0 defined the Wireless Markup Language (WML) specification for building WAP Web sites. Because WML was criticized for its incompatibility with the well-accepted HTML, WAP 2.0 defined XHTML MP (Extensible HTML Mobile Profile), which is a subset of HTML that also conforms to XML syntax. WAP Web sites are specifically designed for viewing in mobile browsers.

### FULL-FEATURED HTML

As the technology of the mobile platform progresses, new mobile browsers such as Apple iPhone Safari, Google Android Browser, and RIM BlackBerry Browser are close to being full-fledged browsers that can be used on desktops. They can support full-featured HTML, CSS, JavaScript, and AJAX, allowing access to most Web pages rather than being limited strictly to WAP Web sites. Since most Web sites built with HTML work on these mobile browsers, organizations don't have to build specific WAP Web sites in addition to their existing "normal" Web sites to accommodate mobile browsers. For Web sites that don't work with these mobile browsers, only the user interface needs to be changed to accommodate the small screen. The back-end business logic can remain mostly unchanged in support of the new user interface.

**Figure 5.9** A simple Web map application, built in PHP/HTML and ArcGIS Server REST API, is shown running on the Apple iPhone, BlackBerry Curve, and Google Android, respectively.

Blackberry, RIM, Research in Motion, SureType, SurePress, and related trademarks, names, and logos are the property of Research in Motion Limited and are registered and/or used in the U.S. and countries around the world. Courtesy of GeoEye and i-Cubed.

## MOBILE BROWSER PLUG-INS

Mobile browser plug-ins, such as Flash Lite, Silverlight Mobile, and JavaFX Mobile, claim to take up only a small amount of memory yet achieve rich applications that will deliver high-quality mobile-optimized media. This approach is still new and so is largely untested at this point.

The main advantages of the mobile browser are that it can reach a broad range of devices (figure 5.9) and that it is easy to develop mobile applications. Developers can use technology to build mobile Web sites that is similar to what they would use for building "normal" Web sites (Shaner and Sai 2008). The main disadvantages of the mobile browser are that it cannot operate when disconnected, its usability is not as rich as native applications, and it cannot directly access peripheral devices such as a GPS receiver. Because of this, the mobile browser does not fit applications where high spatial accuracy is required.

### 5.3.3 MESSAGING APPROACH

Mobile messaging is prevalent on most cell phones. Mobile messaging includes Short Message Service (SMS), which is text-only messaging, and Multimedia Messaging Service (MMS), which can send photos, audio, video, and rich text in addition to basic text. Because SMS is the only data communication method available for many low-end cell phones, a message service can be an effective way to reach a large consumer base.

Some examples of this approach include Yahoo and Google, which allow mobile users to obtain business listings, weather, and directions via a short text message. You can send a search query as a text message. The results will be answered in one or more text messages depending on the length of the result.

In addition to being used by itself, the messaging approach can also be used in combination with other approaches. For example, some Web sites allow you to search for places or create maps and

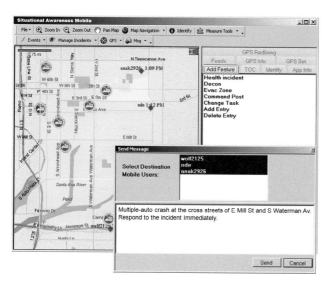

**Figure 5.10** The "group messaging by area" function available in the ESRI Situational Awareness Bundle and Loma Linda University Medical Center AEGIS allows the command center to select an area on the map and send group text messages to field crews within that area.

Courtesy of Tele Atlas North America, Inc.

directions from your desktop computer and send the results to your cell phone via text messages. Another example is the "group messaging by area" function offered in the ESRI Situational Awareness Bundle and the Loma Linda University Medical Center Advanced Emergency GIS (AEGIS) application (see section 5.4.2). Using this application, field crews can report their locations to the server. The command center can then draw an area on the map and send commands as group text messages to field crews within that area (figure 5.10). This application is especially effective for emergency communications since it is often difficult to know who is where and what their phone numbers are.

## 5.4 APPLICATION CASE STUDIES

The advantages of mobile GIS are increasingly being recognized and used in a variety of fields. This section introduces case studies in survey and inspection, damage assessment, emergency response, and LBS.

### 5.4.1 SURVEY, INSPECTION, AND INVENTORY

Traditionally, the processes of field data collecting and editing have been time consuming and prone to error. Field workers would usually take scratch paper or paper maps to the field, sketch on them, and write their notes by hand. Once field workers returned to the office, these field sketches and notes would be deciphered and manually entered into the GIS database. The reentry process is not only cost prohibitive, but also prone to error. The result is that GIS data may not be as up-to-date or accurate as it could have been (ESRI 2007a).

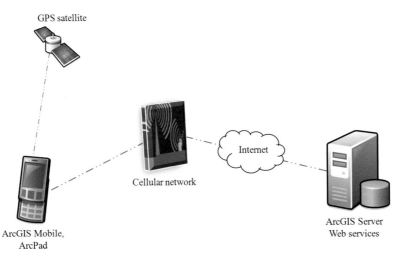

**Figure 5.11**   In wireless communications, field data collected with ArcGIS Mobile can be synchronized immediately with ArcGIS Server.

Mobile GIS has proven to be an ideal solution for overcoming such challenges. It makes it faster and easier for field staff to capture accurate location and attribute information digitally. The use of mobile GIS is not new to field surveys and data collecting. For example, the U.S. Census Bureau has been using ArcPad in various iterations for many of its census activities since 2002.

The recent integration of mobile GIS and the Web has made field surveys and data collecting easier, more accurate, and more efficient. A simplified user interface, enhanced GPS support, and advances in wireless communications have made this possible. Field data can be directly posted to the database server so it is immediately available to other GIS components (figure 5.11), or it can be processed to achieve submeter accuracy before being merged with the master GIS database. For example, Oakland County Animal Control uses ArcGIS Mobile and ArcGIS Server to keep track of pet licenses; the Virginia Department of Forestry developed an Integrated Forest Resource Information System (IFRIS) using ArcGIS Mobile, so that its staff can collect field data on pest infestations and water quality and upload it to ArcGIS Server; and the U.S. Census Bureau built a customized application of ArcPad for the 2010 Census so its field enumerators can follow up on households that don't respond to census workers or whose responses are difficult to read.

Two other case studies illustrate the wide versatility of mobile GIS: Nassau County, Florida, improving its data collecting infrastructure (DCI) with mobile GIS and Victoria, Australia, using mobile GIS to aid in bushfire search operations.

## NASSAU COUNTY, FLORIDA

Nassau County needed to improve its data collecting and ensure department-wide access to infrastructure information. The county also wanted a simple user experience and management process for making mobile deployments so that field staff wouldn't need extensive training to record inspection information.

The county rolled out its DCI for the Utilities Department in a pilot project, where staff collected the locations and relevant information about all fire hydrants in the county's water utility network. The pilot project used a mobile GIS solution based on ArcGIS Mobile, ArcGIS Server, and Trimble GeoXH field devices. The project took the ESRI standard data model for water utilities and built a geodatabase to suit the needs of the county. A map document (MXD) was authored based on the geodatabase, and then published as a Web service. On the mobile client side, the project used the simple task-driven, out-of-the-box ArcGIS Mobile application. With this application, the mobile client would retrieve data and display basemaps served by ArcGIS Server. Data collected in the field was posted to ArcGIS Server via the mobile Web service. The updated data was immediately viewable in a Web map application so the county's GIS mapping specialist could conduct a quality assurance/quality control (QA/QC) process before approving and accepting the field data.

It takes a field team approximately fifteen minutes, on average, to become proficient with this mobile application. Upon using it, field staff become more productive, finishing a greater number of utility location assignments each day and allowing the county to maintain an accurate, up-to-date inventory of all assets in the enterprise. Users are able to quickly download their daily projects; access the specific GIS data they need for the project directly over the wireless network; and upload their work to the server, ensuring that no data is lost. With the field data in digital format, viewable, and able to be queried in a QA/QC map viewer, the county's GIS mapping specialist no longer needs to manually edit separate paper map books for each department. This application can be used for multiple projects involving multiple departments. The Nassau County solution greatly improves the efficiency of the county's data collecting infrastructure and provides a tool for supporting business processes that require fast access to infrastructure maps (ESRI 2009).

## VICTORIA, AUSTRALIA

The Black Saturday bushfires, a series of bushfires that ignited around February 7, 2009, and burned across the Australian state of Victoria during extreme weather conditions, resulted in Australia's highest loss of life from a bushfire. This unprecedented catastrophe covered more than 220,000 hectares, caused a death toll of 173, and injured around 500 people. After the blaze, the Victoria state government required Victoria Police to search every property within the Kinglake Complex fire perimeter for victims and survivors. The mandate required police to inspect each driveway from the road to the house, extending to 50 meters on either side, and to search the perimeter of the residence and outbuildings, extending 50 meters.

This search was an enormous undertaking, considering the large area involved and the legal requirements for documentation. In addition, many towns and communities ceased to exist as a result of the bushfires, making identification of human remains more difficult. Familiar landmarks, such as street signs, mailboxes, residences, and businesses, were reduced to smoldering rubble. These challenges prompted the use of geospatial technologies and real-time mobile GIS to determine the extent of the devastation and document the location of human remains.

Search teams used rugged PDAs running customized ArcPad software for the task (figure 5.12). GIS technology helped police to manage the assignments for each team and track the progress of the search. The NAVTEQ street map helped the field teams navigate to their assignments, and the built-in GPS placed them in the correct parcels. The digital forms in the mobile GIS replaced the paper forms that had been used previously. The use of digital forms ensured that complete and accurate information was recorded, since users could not advance to the next screen until all the required information was entered. The collected information was easily integrated into the existing

**Figure 5.12** Mobile GIS was used to help Australian bushfire search teams navigate to the correct parcels even after street signs were destroyed in the Black Saturday brushfires. The digital forms and GPS-enabled digital cameras were used to document the properties searched.

Victoria Police Web mapping site via a Web map server feed as soon as the mobile devices were synchronized with ArcGIS Server in the Rescue Coordination Center. The Telstra 3G cellular network was used to synchronize the data. Any authorized search manager could tell at a glance which areas had been cleared in the search and where to concentrate resources for the next shift. The use of mobile GIS greatly reduced the length of the search and created accurate means to assess the death toll (Barras 2009).

## 5.4.2 EMERGENCY RESPONSE

Emergency response commonly involves one or more emergency operations centers (EOCs) and dozens to hundreds of field responders. In an emergency situation, there may be only a few short moments to make decisions that can impact the lives of victims and fellow responders. Command centers need the most up-to-date field incident information to make sound decisions. Collaboration and coordination between incident commanders and field responders is crucial. Real-time and server-connected mobile GIS can be used to meet these requirements. Numerous projects employ mobile GIS in emergency response situations, including the following:

- AEGIS (see section 4.3 for project introduction), developed by Loma Linda University Medical Center (LLUMC) jointly with ESRI (Kolbasuk McGee 2008)

- ESRI Situational Awareness Bundle, a ready-to-operate hardware, software, and data solution that provides a geospatial framework for immediate and long-term situational awareness needs

- ALLHAZ (All Hazards Emergency Operations Management System), developed by Jackson State University National Center for Biodefense Communications in partnership with ESRI, provides all field operations personnel with a standardized, scalable, geospatially enabled tool to assist in planning for, mitigating, responding to, and recovering from hazards of all sizes (Matlack, Wesp, and Kehrlein 2007)

Mobile GIS is an important component of these systems because of the wide variety of functions it provides:

- **Field map view and query:** EOCs have GIS servers to provide basemap and imagery services for mobile applications. The mobile application built with the ArcGIS Mobile SDK can dynamically mash up live data feeds, including real-time weather, traffic cameras, resource tracking, hospital status, and active incidents. The information derived from these feeds provides full situational awareness for first responders. Mobile devices are preloaded with the basemap data and critical infrastructure geodatabase so that they can be used even when disconnected from the Internet. Field crews are ensured rapid access to critical data while deploying to and during an incident, even when wireless connection is not available.

- **Resource tracking:** Emergency field crews and vehicles are equipped with mobile phones and GPS receivers so their locations can be reported to the server. The server can track where the field crews, fire trucks, ambulances, and helicopters are via a map display (see section 4.3). Incident commanders can see at a glance where the resources are, determine which resources are closest to a certain incident, and then figure out the best way to dispatch and deploy the available resources.

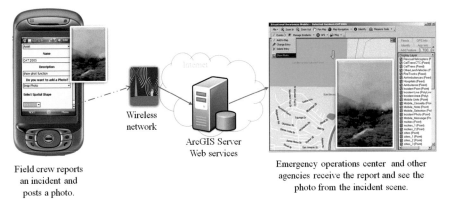

Field crew reports
an incident and
posts a photo.

Wireless
network

ArcGIS Server
Web services

Emergency operations center and other
agencies receive the report and see the
photo from the incident scene.

**Figure 5.13** Mobile GIS allows field crews to collect incident locations, attributes, and multimedia information and report it to the EOC database via wireless network. The EOC and other partner agencies can then use the information to make decisions quickly.

Courtesy of Tele Atlas North America, Inc.

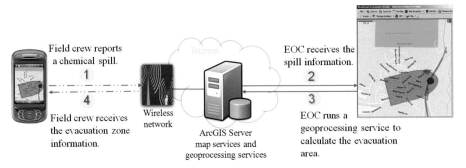

Field crew reports
a chemical spill.

**1**

**4**

Field crew receives
the evacuation zone
information.

Wireless
network

ArcGIS Server
map services and
geoprocessing services

EOC receives the
spill information.

**2**

**3**

EOC runs a
geoprocessing service to
calculate the evacuation
area.

**Figure 5.14** Exchanging data, maps, and geoprocessing capabilities between mobile clients and the GIS server facilitates coordination and collaboration among first responders, the command post, and EOCs.

Courtesy of Tele Atlas North America, Inc.

- **Incident reporting:** Field crews can use the mobile GIS client to collect firsthand information about the location, magnitude, and other details of an incident to compile a timely report. They can also collect photos and send the information to the server in real time via the wireless network. The information is immediately available to the EOC and other agencies. The EOC and other participating agencies can make informed decisions quickly by having quick and accurate updates as the incident unfolds (figure 5.13).

- **Coordination and collaboration:** Mobile GIS can exchange data, maps, analytical results, and text messages with the office and with other mobile users, in support of efficient coordination and collaboration. The group messaging by area function allows the command center to select an area on the map and send text messages to all field crews in the area. This is especially important for large-scale response scenarios when there are many field responders on the scene and it is difficult to locate the phone numbers of individual field crew units. Mobile GIS clients share the same ArcGIS Web services as those in the office. Exchanging data, maps, and analytical results can facilitate coordination of the acts of multiple parties. For example, in a chemical spill, mobile GIS can facilitate the following communication and coordination (figure 5.14):

1. Field crews use mobile GIS to collect and report the spill location, chemical type and quantity, and wind speed and direction to the server.
2. The EOC receives the spill information.
3. The EOC runs a geoprocessing service to calculate the neighborhoods that need to be evacuated.
4. The ArcGIS Server Web service shares evacuation zone information with field crews so they can take action, and with other EOCs so that incident commanders can stay on top of the situation.

### 5.4.3 LOCATION-BASED SERVICES

Location-based services (LBS) have expanded into many aspects of daily life. Instead of using one application as a case study, here is a general series of applications you might need when traveling to a new city:

- **Wayfinding and navigation:** Many mobile map applications can help you find your way to the hotel you reserved—for example, by showing your position on a map or by providing the optimum route for driving, walking, or public transit. On your way, you can check traffic information online (figure 5.15).

- **Querying nearby POIs:** After checking into your hotel, you might be hungry and want to find the closest restaurants. In addition to the previously mentioned mobile applications, you could also use AroundMe or the Yellow Pages mobile applications to find nearby businesses or services. The responses may consist of an ordered list or a simple map centered on your current location. Click a business to get its address, phone number, and links that can help you get there (see figure 1.7).

- **Locating people:** After a meal, you might like to see if there are any friends close by. You can use mobile social mapping services such as Locatrix Communications Uandme, Yahoo Friends on Fire, or Loopt to see where your friends are, or make new ones.

- **Receiving alerts:** Say you have a meeting scheduled at a certain time. A mobile application can calculate how long it will take to get there and when you should leave to get there on time, and remind you ahead of time so you won't be late.

- **Mobile advertising:** After the meeting, perhaps you want to go shopping. You can use mobile applications such as mobiQpons and Coupious to see if there are any local sales or commercial promotions going on. If so, you can show the digital coupons on your phone to the cashiers and save money when you shop there.

These LBS functions are for consumers. But LBS also provides new business opportunities for providers as well. To name a few:

- **Finding someone or something:** finding lost persons, missing pets, lost phones, and lost cars

**Figure 5.15**  Examples of LBS include, from left, viewing real-time traffic using Microsoft Bing Maps Mobile, viewing highway photos with Traffic Vizzion, and finding coupons from nearby businesses with mobiQpons.

Blackberry, RIM, Research in Motion, SureType, SurePress, and related trademarks, names, and logos are the property of Research in Motion Limited and are registered and/or used in the U.S. and countries around the world. Courtesy of Vizzion and MobiQpons, Inc.

- **Resource tracking:** tracking drivers for taxi companies, rental cars for car rental companies, and delivery vehicles for logistics services companies
- **Proximity-based notification:** used for targeted advertising, buddy lists, common profile matching (dating), automatic airport check-in

## 5.5 CHALLENGES AND PROSPECTS

Though great progress has been made in building and applying mobile GIS over the past several years, the technology is still in its infancy. This section discusses issues such as privacy, security, and industry standards that can affect the adoption and development of mobile GIS. It then delves into the opportunities that mobile GIS can offer in the future.

### 5.5.1 PRIVACY, SECURITY, AND STANDARDS

The acceptance of mobile GIS has been hindered, to some extent, by a number of social, technical, and industry factors, including location privacy, security, and compatibility with standards.

#### LOCATION PRIVACY

The location awareness feature of mobile GIS allows the user's current location to be tracked, which can pose a threat to location privacy. For some field employees, such as delivery drivers and utility engineers, having their location monitored is part of their job. But for a large number of users, information about their current location is considered private information, and it is their right to have such information protected. Without proper protection, the information conceivably could be abused or unfairly used for such location-based activities as unsolicited marketing or even harmful encounters such as stalking or physical attacks. Intrusive inferences about an individual's personal information such as personal preferences, state of health, or political views could also be gleaned from this information (Duckham and Kulik 2007).

The different strategies for protecting mobile users' location privacy are summarized by Matt Duckham and Lars Kulik in the following four categories, though they indicate that the technology is not yet sufficiently tested to fully support these strategies:

- **Regulatory strategies:** including notifying the user about who is collecting the information and for what purpose, asking the user's consent, and protecting against unauthorized access, disclosure, and use

- **Privacy policies:** using trust-based mechanisms for proscribing certain uses of location information, such as the P3P (Platform for Privacy Preferences) mechanism developed by W3C

- **Anonymity:** dissociating information about an individual, such as location, from that individual's actual identity

- **Obfuscation:** downgrading the quality of information about a person's location

## SECURITY

Mobile devices store important personal and work information, such as phone contact lists, text messages, e-mail accounts, bank accounts, and sometimes GIS data that is preloaded or just collected. Such valuable information and the large number of owners make mobile devices an attractive target for computer virus attacks. Wireless communications, especially free Wi-Fi, are vulnerable to unwanted intrusions and eavesdropping. Plus, mobile devices are so small that they are easily lost or stolen.

If your mobile device was compromised or stolen, not only could you be receiving expensive telephone bills for calls someone else was making, but more importantly, you could also be giving away sensitive personal or company information. For example, if a customer's contact information was compromised, someone else could be communicating with your customers on your behalf. This could also lead to leaked corporate business data and malicious intrusion into your organization's database and crucial systems.

Measures exist to enhance mobile device security. While not everyone has special modified cell phones with top-notch encryption features like President Obama's BlackBerry, you can still resort to solutions like BlackBerry Enterprise security that enables you to issue an "Erase data and disable handheld" command if your phone was lost or stolen. Other recommendations are similar to those for desktop computers: lock the mobile device with a password, do not install illegal software, and do not open suspicious attachments.

## COMPATIBILITY WITH STANDARDS

Developing mobile GIS that supports all mobile platforms faces several barriers. There are many mobile operating systems, but they offer inconsistent features and SDK APIs. The wide assortment of mobile browsers implements Web standards differently, so Web pages are displayed inconsistently in different browsers. Many countries have implemented 3G networks and have started planning 4G networks, but each country has different standards. Thus, a mobile device that works in one country might not work in another. There are huge digital divides in the wireless data transfer speed among different regions in a country and different countries across the world. With these variations, it is difficult to provide a mobile GIS application that works consistently on all types and all configurations of mobile phones.

Mobile platforms are in a stage of metamorphosis. As each vendor continues to offer special features to gain competitive advantage, it may take several years for standards bodies to catch up with the changes and define standards that all vendors agree on and observe.

## 5.5.2 TOWARD UBIQUITOUS GIS

The future development of mobile GIS can make great contributions to many research disciplines and application areas, including augmented reality, public participation GIS, dynamic demography, and 4D GIS.

### AUGMENTED REALITY

**Augmented reality (AR) combines information from the database with information from the senses** (Longley et al. 2005). With mobile GIS, a user is usually situated within the subject area. Mobile GIS can augment the user's sense of place by providing information about the subject area from the GIS database or from the Web that is not immediately perceptible.

Mobile GIS-aided AR can be applied in many domains such as medical settings, where mobile GIS can replace or enhance the work of senses that are impaired or absent, for example, visually impaired people can use mobile GIS to navigate from place to place. Another example is tourism and sightseeing, where information such as the historic landscape or future design plans for the area at hand can be superimposed on the user's field of view.

### PUBLIC PARTICIPATION GIS

Public participation GIS (PPGIS) refers to bringing the professional practices of GIS into the public domain (see section 10.2). The great penetration of mobile phones provides a GIS platform to a large number of citizens. Imagine each cell phone owner as a moving sensor. Mobile GIS means more than making maps available to the public. It can also create real-time environmental monitoring networks that harvest the collective intelligence of the public at large.

### DYNAMIC DEMOGRAPHY

The demographic information from organizations such as census agencies is generally static. But population is inherently dynamic—for example, population density in different areas of a city changes from morning to evening. Since people leave "digital trails" with their cell phones, it is possible to gauge the spatiotemporal patterns of a population by collecting and analyzing data from cell phone users. Such information can benefit a number of applications, including studies of human behavior, mapping of real-time traffic for streets without traffic sensors, transportation planning, choosing retail store locations, and simulating the spread of an epidemic.

## 4D GIS

The application domain of mobile GIS is usually wherever and whenever geospatial events are taking place—that is, at the moment, and outside the office. Mobile GIS can record, report, and analyze rapidly changing phenomena in situ. This dynamic and real-time capability of mobile GIS embraces the dimension of time and helps extend GIS from 3D (X, Y, and Z) to 4D (X, Y, Z, and time) (Drummond, Joao, and Billen 2007).

The number of cell phone subscribers worldwide is going to continue to increase—up to 5 billion in 2010 (ITU 2010). With pre-4G services already available on the market, real 4G is right around the corner. Future mobile phones and ultra-mobile PCs promise exciting advances in computer technology, including the increased use of mobile GIS. As GIS is integrated with a multitude of computing devices and connected to the Web with the faster 4G network, GIS will become ubiquitous. The omnipresence of GIS will open the door to making mobile GIS available to anyone—for anything, anytime, anywhere.

---

## Study questions

1. What is mobile GIS? What advantages does it have?

2. Describe the differences between consumer and enterprise mobile GIS applications.

3. What are the available mobile positioning technologies? Explain their pros and cons.

4. What technical challenges face mobile GIS applications? How do these challenges affect mobile GIS application development?

5. What are the available options for mobile GIS application development? If an application requires high location accuracy and disconnected operation, what solutions and products would you choose? Why?

6. How can mobile GIS assist emergency response operations?

7. What is LBS? How is it (or can it be) used in your daily life or work?

8. What is location privacy? What are the strategies to protect it?

9. Name research areas and applications where mobile GIS can contribute.

## References

Barras, Greg. 2009. Mobile GIS aids Victoria bushfires search operations. *ArcNews* (Summer). `http://www.esri.com/news/arcnews/summer09articles/mobile-gis-aids.html` (accessed August 9, 2009).

B'Far, Reza. 2005. *Mobile computing principles: Designing and developing mobile applications with UML and XML.* Cambridge: Cambridge University Press.

Drummond, Jane, Elsa Joao, and Roland Billen. 2007. Current and future trends in dynamic and mobile GIS. In *Dynamic and mobile GIS: Investigating changes in space and time,* ed. Jane Drummond, Billen Roland, Elsa Joao, and David Forrest. Boca Raton, Fla.: CRC Press/Taylor & Francis.

Duckham, Matt, and Lars Kulik. 2007. Location privacy and location-aware computing. In *Dynamic and mobile GIS: Investigating changes in space and time,* ed. Jane Drummond, Roland Billen, Elsa Joao, and David Forrest. Boca Raton, Fla.: CRC Press/Taylor & Francis.

ESRI. 2007a. ArcGIS Server and mobile GIS. Infrastructure Management Series. `http://www.esri.com/industries/public-works/business/infrastructure_mngment_series-3.html` (accessed June 8, 2009).

———. 2007b. Mobile GIS best practices. `http://www.esri.com/library/bestpractices/mobile-gis.pdf` (accessed June 5, 2009).

———. 2009. Nassau County mobile GIS improves data collection infrastructure. `http://www.esri.com/library/casestudies/nassau-county.pdf` (accessed August 12, 2009).

International Telecommunication Union. 2009. Measuring the Information Society: The ICT development index. `http://www.itu.int/ITU-D/ict/publications/idi/2009/material/IDI2009_w5.pdf` (accessed August 11, 2009).

———. 2010. ITU sees 5 billion mobile subscriptions globally in 2010. `http://www.itu.int/newsroom/press-releases/2010/06.html` (accessed May 28, 2010).

Kapoor, Neeraj. 2008. Mobile architecture for enterprises. One Associates Technologies. `http://www.one-associates.com/Enterprise_Mobile_Architecture_v1.0.pdf` (accessed June 2, 2009).

Kolbasuk McGee, Marianne. 2008. Hospital implements high-tech emergency response system. *InformationWeek.* `http://www.informationweek.com/news/personal_tech/gps/showArticle.jhtml?articleID=208800864` (accessed January 28, 2010).

Kowoma. 2009. Sources of errors in GPS. `http://www.kowoma.de/en/gps/errors.htm` (accessed May 31, 2009).

Lange, Jian, Myles Sutherland, and Shane Clarke. 2009. ESRI mobile GIS solutions overview. ESRI International User Conference, San Diego, Calif., July 13–17.

Longley, Paul A., Michael F. Goodchild, David J. Maguire, and David W. Rhind. 2005. *Geographic Information Systems and Science.* 2nd ed. San Francisco: John Wiley & Sons.

Maguire, David. 2007. The changing technology of space and time. In *Dynamic and mobile GIS: Investigating changes in space and time,* ed. Jane Drummond, Roland Billen, Elsa Joao, and David Forrest. Boca Raton, Fla.: CRC Press/Taylor & Francis.

Matlack, Elizabeth, Tom Wesp, and Dave Kehrlein. 2007. ALLHAZ field-level emergency operations concept. ESRI Homeland Security GIS Summit, Denver, Colo., November 5–7. `http://proceedings.esri.com/library/userconf/hss07/docs/allhaz_workshop.pdf` (accessed June 4, 2009).

*Popular Science.* 2002. Your phone knows where you are. `http://www.popsci.com/scitech/article/2002-06/your-phone-knows-where-you-are` (accessed June 15, 2009).

Shaner, Jeff, and Jayant Sai. 2008. The mobile Web. *ArcGIS Mobile Blog.* `http://blogs.esri.com/Dev/blogs/mobilecentral/default.aspx` (accessed June 3, 2009).

Spiekermann, Sarah, and Humboldt-Universitat zu Berlin. 2004. General aspects of location-based services. In *Location-based services,* ed. Jochen Schiller and Agnes Voisard. Oxford: Elsevier.

Sutherland, Myles, and Michael Miller. 2009. Implementing ArcGIS Mobile. ESRI Developer Summit, Palm Springs, Calif., March 23–26.

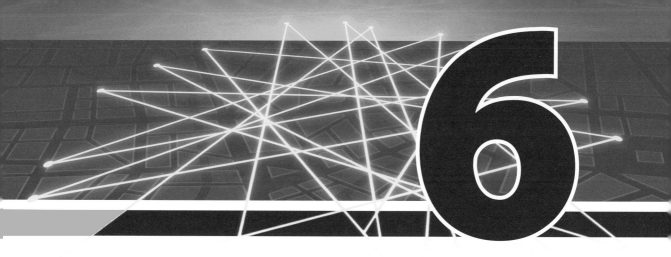

# GEOPORTALS

People and organizations have always exchanged geographic information, but the practice has grown exponentially in recent years with the popularization of the Internet and the Web. An increased amount of geospatial data is being collected by virtue of more user-generated content on the Web, in addition to data from satellites, ground and ocean sensors, and GPS units (Goodchild, Fu, and Rich 2007). Geoportals—that is, Web sites where geospatial information can be discovered—make it easier for users to find, access, and use geospatial information. Geoportals emerged in the mid-1990s, and many countries have developed geoportals as an important part of their SDI (spatial data infrastructure).

It is important to introduce geoportals in the context of Web GIS, because geoportals are a type of Web GIS application. They facilitate geospatial information sharing and provide a resource for developing other Web GIS applications.

This chapter has four sections. Section 6.1 introduces the concept of geoportals and their importance to SDIs. Section 6.2 presents geoportal functionality and the geospatial metadata behind the functions, and then discusses the distributed and centralized catalog architectures. Section 6.3 uses case studies to illustrate the characteristics and design considerations of different types of geoportals. Section 6.4 discusses challenges such as metadata complexity, semantic search, digital rights management, and data mining, as well as the prospects for geoportals.

## 6.1 CONCEPT AND USES

The word "portal" means gate or entrance. It was adopted in the mid-1990s to form new terms such as Web portals, Web sites that serve as the gateway to other Web sites, and geoportals, gateways to geospatial information.

### 6.1.1 GATEWAYS TO GEOSPATIAL INFORMATION

Derived from the Latin word *porta*, indicating a doorway, **a Web portal, or simply portal, is a Web site that functions as an entry point to the World Wide Web. A portal provides search tools that help users find information on the Web, and it usually also provides categorized links to many online resources.** Web portals have greatly expanded since the mid-1990s, as the volume of content available over the Web has grown exponentially and Web users have sought ways to quickly find the content they want. Web portals provide an easy, direct means of indexing, ordering, and displaying an otherwise overwhelming amount of information. You can visit Web portals such as Google, Microsoft Bing, and Yahoo; enter a simple word or phrase or navigate through a hierarchical directory; and quickly find relevant sites from the billions of Web pages available on the Internet. Web portals saved the Web from being bogged down in a sea of data and have played a significant role in the success and popularity of the Web (Tang and Selwood 2005).

Web portals can be classified by their range of contents into general (or horizontal) portals versus specialized (or vertical) portals. A geoportal, as indicated by the prefix "geo," is a portal that specializes in geospatial information. **A geoportal, also referred to as a spatial portal, is a Web site that provides a single point of access to geospatial data, Web services, and other geospatially related resources. Put more simply, a geoportal is a Web site where geospatial resources can be discovered** (Tait 2005). Geoportals, as gateways to geospatial information, facilitate geospatial information sharing between providers, who own the information, and users, who need the information. Geospatial information providers publish the metadata—that is, the

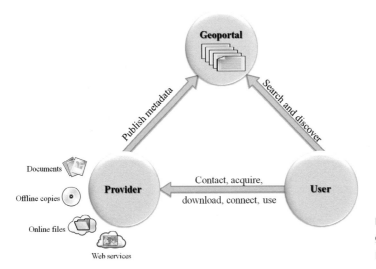

**Figure 6.1** Geoportals form a gateway between geospatial resource providers and users.

description of the geospatial data, Web services, and documents—to geoportals. Users can browse or search through geoportals to find relevant geospatial resources and evaluate their applicability. Users can then download the online data, connect to and use Web services, or contact the content provider for offline data or documents (figure 6.1).

Geoportals are unique from general-use Web portals in that they specialize in discovering geospatial contents. Geoportals enable users to search for contents not only with keywords, but also with the spatial and temporal dimensions of information—for example, search based on a user-defined spatial area or moment in time. This capability is of particular importance to GIS users.

The Alexandria Digital Library (ADL) and U.S. National Geospatial Data Clearinghouse (NGDC) (see section 6.2.3) are among the early examples of operational geoportals. ADL is an online geolibrary with a goal of providing a comprehensive range of library services for spatially indexed geographic information, including digitized maps, images, and textual materials. Users are able to retrieve materials from the library on the basis of information content as well as by reference to spatial location (Smith and Frew 1995; Goodchild 2004).

## 6.1.2 THE "FACE" OF SDI

Geographic information is vital to making sound decisions at the local, regional, national, and global levels. It is crucial in terms of addressing social, economic, and environmental issues of pressing importance. Yet while geospatial information has been collected by numerous organizations, many are not motivated to share their data with others. This creates isolated islands of information, which can make it difficult for organizations to discover and obtain the data they need. This, in turn, leads to underused data, duplicate data collecting, wasted funding, time delays, and inadequate decision making (Sun and Shi 2002; Wang 2005).

SDI (see chapter 7), as a concept, is charged with addressing these issues. SDI is basically the technology, policies, standards, and human resources necessary to acquire, process, store, distribute, and improve the use of geospatial data. SDI can be at the global level (GSDI), national level (NSDI), regional level, or local level. The primary objective of SDI is to maximize the overall use of geographic information that is held by a wide range of stakeholders in both the public and private sectors.

Because geoportals help to realize the objectives of SDI to facilitate collaboration, they have been developed as a major part of SDI. Without geoportals, it can take an organization months to make phone calls, write letters and e-mails, and physically travel to a place to inquire about the existence of relevant geospatial data held by another agency, and then to verify whether that data addresses the intended use. With geoportals, the search time can be reduced from days to seconds. As the broker between the data provider and the user, **geoportals are an important and highly visible component of SDI, serving as the "face" of SDI** (Tang and Selwood 2005; Hogeweg 2009a). In the past decade, more than 100 countries (see following examples) have embarked on some form of SDI initiative (Masser 2007; Li 2006; Maguire and Longley 2005).

The United States is among the countries that got a head start on developing SDI. The United States, in 1990, established the Federal Geographic Data Committee (FGDC, `http://www.fgdc.gov`), an agency consisting of the U.S. Geological Survey (USGS) and members of all major federal departments with an interest in geospatial information. The FGDC promotes the coordinated development, use, sharing, and dissemination of geospatial data. Federal, state, and local governments, academia, and the private sector have actively participated in building SDIs following FGDC guidelines.

- In the 1990s, the FGDC and USGS coordinated the establishment of the National Geospatial Data Clearinghouse, which was retired circa 2003. The NGDC was a major geoportal for geospatial datasets and was one of the most obvious SDI success stories.

- In 2002, the Geospatial One-Stop (GOS, `http://www.geodata.gov`) Portal (figure 6.2) was initiated as one of President Bush's twenty-four e-government projects. GOS is sponsored by the federal Office of Management and Budget (OMB) and is coordinated by USGS and the FGDC. GOS enhanced the NGDC geoportal with many new features such as a centralized architecture, improved performance, and support for a range of open standards. GOS replaced the NGDC portal as the main source of geospatial information circa 2003, and it holds about 430,000 records of geospatial content as of May 2010 (see section 6.3.3).

- In 2009, as a priority of President Obama's Open Government Initiative, the U.S. government built Data.gov to increase the ability of the public to find, download, and use datasets that are generated and held by the federal government. Although it is a general Web portal, geospatial data is the main content, and its geospatial data catalog search is supported by GOS behind the scenes.

- Many other geoportals exist in the government, academic, and private sectors. NASA's Global Change Master Directory (GCMD, `http://gcmd.nasa.gov`), formerly called the NASA Master Directory, gives users access to Earth science datasets and to services that are relevant to global change and Earth science research. Other geoportals include the Mississippi Geospatial Clearinghouse, Arkansas GeoStor, Kentucky Watershed Modeling Information Portal (KWMIP), and many others. In the private sector, ESRI developed the Geography Network geoportal in 2000, a big step toward Web services discovery and dynamic composition. The Geography Network has been replaced by ArcGIS Online and ArcGIS.com, which provide enhanced capabilities and convenience (see section 6.3). ESRI also developed the ArcGIS Server Geoportal Extension product (initially named GIS Portal Toolkit), which is used to support many of the geoportals mentioned here.

In Canada, the GeoConnections program (`http://www.geoconnections.org`), an important component of the Canadian Geospatial Data Infrastructure (CGDI), developed a geoportal that provides an online catalog of a wealth of Canadian geospatial information. The geoportal helps users to share geospatial information in four priority areas: public safety, public health, aboriginal community affairs, and the environment and sustainable development.

The Australia and New Zealand Land Information Council (ANZLIC) led the implementation of the Australian Spatial Data Infrastructure (ASDI). A key component of the ASDI is the Australian Spatial Data Directory (ASDD, `http://asdd.ga.gov.au/`), which is a geoportal that provides search interfaces to geospatial datasets throughout Australia.

Many countries in Europe are in the process of developing SDIs in response to the Geographic Information 2000 initiative launched by the European Commission in 1996 and the Infrastructure for Spatial Information in Europe (INSPIRE) initiative launched in 2001. INSPIRE aims to build

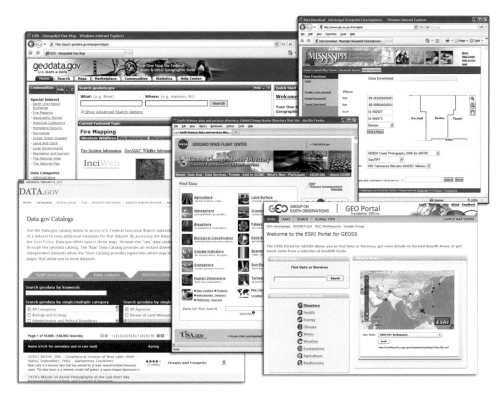

**Figure 6.2**   Example geoportals include (top row) Geospatial One-Stop and Mississippi Geospatial Clearinghouse, and (bottom row) Data.gov, NASA Global Change Master Directory, and GEO (Group on Earth Observation) Geoportal.

Courtesy of U.S. Geological Survey, NASA, MSITS, and GEO.

a European SDI that goes beyond national boundaries. INSPIRE later became a directive that went into force in 2007. INSPIRE will be implemented in various stages, with full implementation required by law by 2019. The INSPIRE directive requires the European Commission to establish a community geoportal where member states will provide access to their infrastructures through this main geoportal as well as through any access points they decide to operate and maintain. This geoportal is one of the most ambitious geoportal projects in the world, since it will have to address the differences in language, business practices, and cultural affairs of its members. A prototype INSPIRE geoportal was released around 2003 with the goal of identifying issues related to its implementation, development, and operation (Bernard et al. 2005).

The National Geospatial Information Coordination Committee (NGICC) and a number of other agencies are coordinating the development of SDI in China. China initiated a series of science data-sharing pilot projects in 2001. A number of the pilot projects, including the Data Sharing Infrastructure of Earth System Science (`http://www.geodata.cn`), provide geoportals that have led to great progress in sharing earth science data and basemap data across a broad sector.

Along with the development of SDIs in many countries, international organizations, including the Global Spatial Data Infrastructure Association (GSDIA) and Group on Earth Observations (GEO), are working to promote the development of GSDI. The SDI "cookbook" (Nebert 2004) by GSDIA has been widely used to guide building SDIs around the world. The Global Earth Observation System of Systems (GEOSS) Geoportal (http://geoss.esri.com/geoportal) by GEO provides scientists with easy access to Earth observation data and Web mapping services.

## 6.2 FUNCTIONS AND ARCHITECTURES

Geoportals provide an array of functionality to administrators, content providers (publishers), and users. Fundamental to their operation is the metadata catalog, which can be managed in a distributed or a centralized architecture.

### 6.2.1 GEOPORTAL FUNCTIONS

Geoportals should provide typical functionality suitable to each of the three roles—publisher, administrator, and user—associated with a geoportal (figure 6.3).

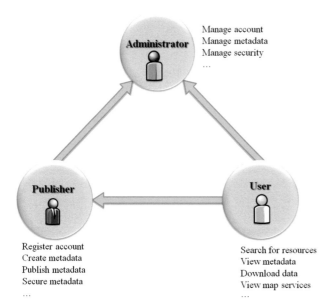

Manage account
Manage metadata
Manage security
...

**Administrator**

**Publisher**

**User**

Register account
Create metadata
Publish metadata
Secure metadata
...

Search for resources
View metadata
Download data
View map services
...

**Figure 6.3** Basic functionality provided by geoportals is divided among publishers, administrators, and users.

## PUBLISHERS

Publishers provide content to the geoportal. Geoportals typically offer publishers the following functionality:

- **Creating accounts:** allows publishers to sign up, become a publisher, and join certain groups
- **Creating metadata:** helps publishers to create metadata that typically complies with certain standards
- **Publishing metadata:** helps publishers contribute metadata to the portal catalog
- **Securing metadata:** allows publishers to specify whether their metadata is public or only accessible by certain groups
- **Other functions:** includes registering metadata repositories to be harvested. This is usually for publishers who have their own geoportals and would like to have their contents automatically included in another geoportal (see sections 6.2.3 and 6.3.3).

## ADMINISTRATORS

Administrators manage content in the geoportal. Geoportals typically offer administrators the following functionality:

- **Managing accounts:** allows administrators to review accounts, approve or disapprove requests to be publishers, and to join certain groups
- **Managing metadata:** allows administrators to review metadata records that are published and approve or disapprove them
- **Managing security:** assists administrators in reviewing the privileges of each group and configuring the contents that each group can access
- **Other functions:** includes managing harvesting and ensuring that harvesting occurs as scheduled

## USERS

Users search for contents that meet their needs. Geoportals typically offer users the following functionality:

- **Searching metadata:** provides an interface for users to specify criteria such as what (keywords, content type, or format), where (spatial extent, location information), and when (temporal range), and then searches the metadata catalog and returns the matching records in a list sorted by relevance or other properties
- **Viewing metadata:** allows users to view metadata discovered in its original format or transformed formats, which are usually easier to read
- **Downloading data:** provides links to allow users to download the resource just discovered
- **Viewing map services and using Web services:** allows users to preview individual map services or create composite maps with the use of a map viewer

- **Other functions:** includes creating accounts, managing personal profiles, saving certain search criteria, saving created maps, creating composite maps, and notifying users when relevant metadata is published or added to the portal

Geoportal functionality can be provided in a variety of ways (figure 6.4). For example, ArcGIS Server Geoportal Extension can perform a search in the following ways:

- **Geoportal Web site:** Users visit the Geoportal Web site to specify search criteria and review the search results.

- **Web browser quick search box:** Users can register their geoportal as a search provider in Web browsers such as Firefox and Internet Explorer, and then use the quick search box without launching the geoportal Web page.

**Figure 6.4** ArcGIS Server Geoportal provides search capabilities via (A) its Web site, (B) a Web browser quick search box, (C) REST and CSW API, (D) a widget in its Flex-based Map Viewer, and (E) plug-ins for ArcGIS Desktop and ArcGIS Explorer.

Courtesy of NOAA and GEO.

- **Plug-ins and widgets:** Users can install geoportal plug-ins for ArcGIS Explorer and ArcGIS Desktop and widgets for HTML (see figure 4.10A) and ArcGIS Flex Map Viewer, and then search for geospatial resources and use the discovered resources directly in these clients.

- **Programming interfaces:** ArcGIS Server Geoportal Extension provides a REST interface and a CSW interface, which open an array of possibilities to integrate with other application environments. The REST interface can return search result in GeoRSS, KML, HTML, HTML fragments, and JSON formats. These formats support a variety of clients—for example, you can subscribe to GeoRSS via Microsoft Office Outlook and other RSS readers, and HTML can be indexed by general-use Web portals.

## 6.2.2 METADATA STANDARDS AND METADATA 2.0

Metadata, or "data about data," is information that describes other data. **Geospatial metadata refers to the information that describes geospatial data, Web services, or other geospatial resources.**

### IMPORTANCE TO GEOPORTALS

Geospatial metadata, usually presented in XML format, represents the "who, what, when, where, why, and how" of a piece of geospatial data or other resource (FGDC 2006). There is object-level metadata, which describes a single entity, and collection-level metadata, which describes a series or a group of entities (Goodchild and Zhou 2003). Geospatial metadata typically includes the following:

- **Identification,** such as dataset title, citation, abstract, purposes, and keywords

- **Quality,** such as positional accuracy, data completeness, and consistency

- **Spatial reference,** such as the coordinate system and spatial extent

- **Temporal information,** such as the date the data was acquired and the length of time the data is valid

- **Distribution information,** including the distributor and options for obtaining the dataset, such as the format of the resource and the URL used to download the data or access Web services

Metadata is useful in data archiving, assessment, management, discovery, transfer, and distribution. In the geoportal context, metadata has the following uses (Longley et al. 2005; Van Oosterom and Zlatanova 2008):

- **Discovery of resources:** Metadata has a setup that is similar to a library catalog, which organizes the library's books by author, title, and subject so that you can find the books you want. Geospatial metadata includes properties such as geospatial extent, temporal coverage, and resource type. This enables you to specify search criteria that goes beyond author, title, and keywords. Geoportals can search the metadata catalog and return matching records as well as details on how to access the geospatial resources or whom to contact to obtain the resources.

- **Evaluation of resources:** Once you discover a resource, you need to determine whether it will fit a given use—for example, does the dataset or Web service have sufficient spatial resolution and acceptable quality to meet your project requirements? Does it have any restrictions on usage? You can get answers to these questions from the discovered metadata.

- **Use of resources:** Metadata typically includes technical specifications, such as the dataset volume, data format, type of Web service, and software tools needed to handle the resources. Such information will let you use the resources effectively.

- **Contract between the user and the provider (publisher):** Providers can disclose the terms of use in the metadata, such as the limit of Web service uses per day, the right to discontinue the service, copyright attribution requirements, and limitations of liability. When there are disputes or copyright infringement, the data provider can avoid legal responsibilities or claim copyright.

### GEOSPATIAL METADATA STANDARDS

Just as book cards in a library catalog have consistent content items, metadata in a geoportal should also have a consistent format. This enables a geoportal to automatically process metadata easily and support searches based on certain criteria. While some organizations choose to define their own metadata formats for use in their own geoportals, geoportals typically follow certain national or international standards (figure 6.5). **A metadata standard is a common set of terms and definitions that describe geospatial resources.** A metadata standard outlines the properties to be recorded, as well as the values allowed for each property. Metadata standards are the basis for metadata interoperability, so that metadata can be understood within a geoportal and between geoportals.

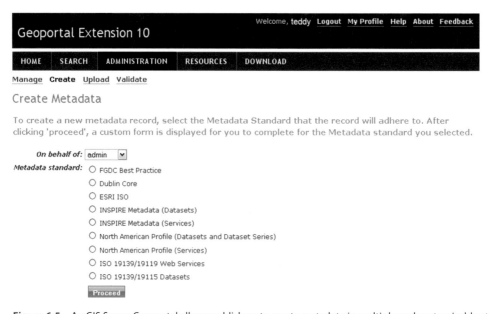

**Figure 6.5**   ArcGIS Server Geoportal allows publishers to create metadata in multiple and customizable standards.

Some of the main geospatial metadata standards include those discussed in the following sections.

### DUBLIN CORE

The Dublin Core metadata is a simple standard for describing cross-domain information resources. Dublin refers to Dublin, Ohio, where the work originated in 1995. Dublin Core was adopted by the International Organization for Standardization (ISO) as ISO Standard 15836. Its greatest advantage is its simplicity and ease of use. It has two levels: simple and qualified. Dublin Core Simple consists of only fifteen elements, such as title, subject, description, type, creator, date, and format, in three groups. Dublin Core Qualified comprises three additional elements (audience, provenance, and rights holder) and a group of element qualifiers that refine the semantics of the elements. Dublin Core is the most popular metadata standard and has been widely used to describe many types of resources, including geospatial resources. However, because of its limited elements, it often needs to be extended with Resource Description Framework (RDF).

### FGDC METADATA (CSDGM)

CSDGM (Content Standard for Digital Geospatial Metadata), commonly known as the FGDC Metadata Standard, is the U.S. federal metadata standard. This standard was drafted in 1992. President Clinton's Executive Order 12096 of 1995 ordered all federal agencies to use this standard to document geospatial data. The standard has been widely implemented by all levels of government and nongovernment organizations. CSDGM includes a hierarchy of data elements and compound elements. It includes seven main sections—identification information, data quality information, spatial data organization information, spatial reference information, entity and attribute information, distribution information, and metadata reference information—and more than 400 elements in total.

While CSDGM is still widely used in the United States, it is important to note that the United States is in the process of adopting the North American Profile of ISO metadata standards discussed as follows.

### ISO METADATA STANDARDS

The international community, through ISO/TC 211 (Technical Committee 211), has developed and approved a series of international metadata standards. ISO 19115 (2003) defines the schema required for describing geographic resources. It has more than 400 elements in its data dictionary, which are divided into fourteen metadata packages (table 6.1). ISO 19139 (2007) provides a common XML specification for describing, validating, and exchanging geographic metadata. It is designed to promote metadata interoperability and exploit ISO 19115's advantages in a concrete implementation specification.

ISO metadata standards are being increasingly adopted. Some organizations are in the process of recasting their previously used metadata standards as ISO metadata "profiles." An ISO metadata profile contains a select set of metadata elements from ISO standards, with the optional inclusion of additional metadata elements as formal extensions of ISO standards. For example, the United

States and other North American countries defined the North American Profile based on ISO 19115 (dataset) and ISO 19119 (services).

| PACKAGE | XML ELEMENT |
|---|---|
| Metadata entity set information | MD_Metadata |
| Identification information | MD_Identification |
| Constraint information | MD_Constraints |
| Data quality information | DQ_DataQuality |
| Maintenance information | MD_MaintenanceInformation |
| Spatial representation information | MD_SpatialRepresentation |
| Reference system information | MD_ReferenceSystem |
| Content information | MD_ContentInformation |
| Portrayal catalog information | MD_PortrayalCatalogueReference |
| Distribution information | MD_Distribution |
| Metadata extension information | MD_MetadataExtensionInformation |
| Application schema information | MD_ApplicationSchemaInformation |
| Extent information | EX_Extent |
| Citation and responsible party information | CI_Citation and CI_ResponsibleParty |

**Table 6.1**   ISO 19115 packages and their main corresponding XML elements

## METADATA 2.0

Most metadata standards are complex since they must account for all kinds of geospatial resources. While there are tools such as ArcGIS Desktop that help data owners create metadata that complies with ISO and other standards, creating standard compliant metadata remains a burdensome task. Many publishers struggle with it.

With the pervasive bottom-up information product phenomena in the era of Web 2.0, content sharing is no longer limited to the scientific and professional community. Everyone who is willing to share their content should be able to do so. With the increased ease of sharing comes the demand for an increased ease in describing the item to be shared. This leads to the need for metadata 2.0, which should be user-centric, easy for contributors to create, and easy for users to understand (Goodchild 2008; Hogeweg 2009b).

The recent popularity of tags represents a move toward metadata 2.0. A tag is basically a keyword or term that describes an item, such as a photo, a video, or a tool. For example, Flickr and YouTube allow you to add free-form tags to each of your pictures or videos—in essence, constructing flexible and easy metadata that make the pictures and videos searchable. Tags can be considered a "bottom-up" type of classification, compared with the traditional hierarchical system (taxonomy) of metadata standards, which are "top-down." Tags allow contributors to quickly share information. Such metadata is often informal and incomplete, from the viewpoint of professionals. Yet tags let users quickly evaluate resources to see if it's what they're looking for. The information gleaned from previewing photo icons or skimming through videos can offset the lack of complete metadata. On many Web sites, the opinions of others about a resource become an integral part of its metadata—albeit, informal metadata. For example, customer reviews about the products available in an online store can become helpful metadata for other customers searching for a particular item.

Different types of metadata are needed for different uses. While metadata 1.0 is still important in enabling professionals and organizations to document their GIS information assets, and to discover and evaluate the applicability of assets for certain uses, metadata 2.0, such as tags, encourages broad participation in the sharing of information and is a strategy that can be adopted by geoportals for many situations.

## 6.2.3 DISTRIBUTED VERSUS CENTRALIZED CATALOGS

Geoportals, especially those at the national and regional levels, typically involve many partners who have metadata to contribute. Depending on whether a geoportal has a centralized metadata catalog or a distributed catalog that accesses multiple catalogs in different locations, the geoportal can be categorized as having centralized or distributed architecture.

### DISTRIBUTED CATALOG

A geoportal with a distributed catalog needs to perform distributed, or federated, searches to fulfill a user's search request by searching multiple catalogs (figure 6.6). One example of a distributed catalog is the NGDC, which later was replaced by GOS. Although initially targeted to federal

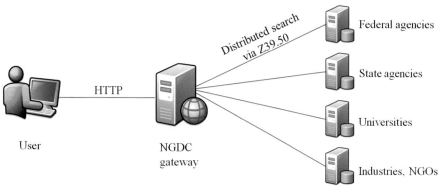

**Figure 6.6** Geoportals with distributed catalogs—for example, the now defunct U.S. National Geographic Data Clearinghouse—need to distribute users' search requests to the participating catalogs, merge the results from each catalog, and then present the merged results to the user.

Clearinghouse nodes at organizations that provide metadata

agencies, the NGDC evolved to include several hundred federal, state, university, and vendor participants in the United States and abroad. The NGDC was set up to help users find spatial data held by participating partners.

Anticipating that there would be hundreds of participating organizations, the NGDC chose a distributed architecture. In this type of architecture, each participant maintains its own metadata catalog server, which is called a "node." The nodes were registered to an FGDC gateway database. The NGDC provided a gateway server, which had an entry Web page that allowed users to specify search criteria, select one or more nodes, and perform a search. The NGDC gateway would then establish connections between the gateway and the selected nodes, pass the metadata query to those nodes, receive the query results from the nodes, merge the search results, and present the merged results to the user. The user could then view the matched metadata records.

The NGDC was one of the most visible success stories in SDI. It partnered with hundreds of organizations that contributed both metadata and data and became a major source of geospatial data. The NGDC was one of the primary data dissemination mechanisms in the United States. However, it also had the limitations inherent in distributed searches, including the following:

- **Difficulty in conducting a global search:** A user has to select from the hundreds of nodes available to perform a search. What a user is looking for could be held by a node that is not chosen.

- **Limitations in performance:** When more than one node is selected, the gateway has to wait until the slowest node responds before it can merge the results from each node.

- **Difficulty in sorting the results consistently:** When the results from different nodes are merged, it can be difficult to sort the results consistently. This sorting has to deal with the different matching and ranking algorithms used by different nodes. What scores high in one node may score low in another. It can be challenging to merge the two results, and then sort them according to ranking.

- **High requirements for participant providers:** Each provider organization needs to have professional staff to build its metadata catalog and set up the related software that supports periodic distributed search requests.

## CENTRALIZED CATALOG

With the advances in computer, database, and network technology, the centralized catalog is becoming popular. A geoportal with a centralized catalog has all its metadata records contained in one catalog. Small and medium-size geoportals can make use of centralized architecture fairly easily. For large geoportals that involve multiple providers, each with its own metadata catalog, a centralized catalog can also be a good strategy. Each of the participant's catalogs can be loaded into the central catalog through metadata harvesting, or synchronization. Each provider can register his or her own catalog and the frequency with which it should be harvested. The ArcGIS Server Geoportal Extension synchronizer (figure 6.7), which operates in many ways like the spiders or crawlers of general Web portals, automatically connects to a registered catalog, retrieves all its metadata on the first visit, and on subsequent visits retrieves only updated data. The ArcGIS Server Geoportal Extension can harvest from many different catalog types such as OGC Catalog Service for Web (CSW), XMLs under Web-accessible folders (WAF), the ESRI metadata service, and the

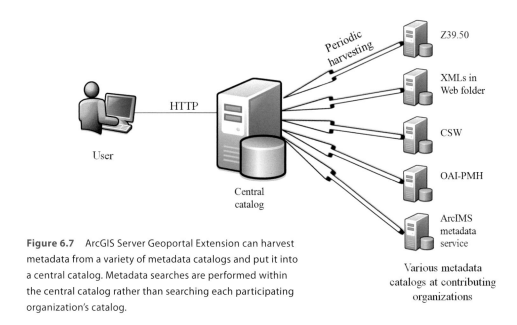

**Figure 6.7** ArcGIS Server Geoportal Extension can harvest metadata from a variety of metadata catalogs and put it into a central catalog. Metadata searches are performed within the central catalog rather than searching each participating organization's catalog.

Open Archive Initiative (OAI). Such a mechanism has been applied to many geoportals, including GOS (see section 6.3.3) and the U.S. EPA GeoData Gateway.

The catalog of a centralized geoportal can grow quite large. To store and maintain such a large volume of metadata records and achieve fast performance, a central catalog may potentially need a lot of hardware, software, and related professional resources. Geoportals with centralized catalogs, however, tend to overcome the limitations inherent in distributed catalogs and yield the following advantages:

- **Easy to conduct global searches:** Each search is against all metadata records in a single catalog database. Users do not need to select nodes.

- **Easy to sort the results consistently:** The search result is from one database and can be easily sorted globally, by relevance or by other properties.

- **High performance:** Since all searches rely only on the central catalog, there is no need to wait for distributed nodes to respond. Because of advances in information technology, a well-designed and well-tuned central catalog, even with a huge volume of records, can still achieve high performance.

- **Low requirements for participant providers:** Providers do not need to have their own geoportals. Their metadata can be directly published into the geoportal via an online metadata form, metadata XML upload, or metadata harvesting of XMLs in Web folders.

At the individual project level, a project such as GOS can choose centralized architecture. However, from a user's perspective, there is always more than one "one-stop" source to search for geographic information—for example, global climate change research data at the NASA Global Change Master Directory, weather data at the NOAA National Weather Service Web site, European data at the

INSPIRE portal, and China data at the Data Sharing Infrastructure of Earth System Science portal as well as other geoportals in China. These are essentially distributed catalogs from the user's perspective. In an ideal world, one source would come to dominate by harvesting catalogs from all other sources, but this would require almost unlimited resources and would require the collaboration of every provider (Goodchild, Fu, and Rich 2007). **Thus, distributed architecture and centralized architecture complement each other and are both needed to share geospatial information.**

## 6.3 GEOPORTAL CASE STUDIES

Geoportals can be classified by their usage, theme, or content format. For example, geoportals can be categorized as personal, departmental, organizational, national, or global geoportals, depending on their usage. By theme, geoportals can be general portals or have more specific orientations such as climate, transportation, emergency response, or other subjects. By format, geoportals range from only online data to only Web services to other formats and combinations. Different types of geoportals have different characteristics and different design considerations (table 6.2). This section presents three case studies, examining different geoportal designs.

| USAGE | PERSONAL AND DEPARTMENTAL GEOPORTALS | ORGANIZATIONAL AND COMMUNITY GEOPORTALS | REGIONAL, NATIONAL, AND GLOBAL GEOPORTALS |
|---|---|---|---|
| **CHARACTERISTICS** | Convenience and seamless integration with the proprietary tools that the individual, organization, or community uses are the focus. National and international standards are less important. | | National and international standards such as ISO metadata and OGC Web services standards are more important. |
| **METADATA** | Simple, for specific purposes; may not comply with standards. | | Important to comply with national or international standards. |
| **PUBLISHING METHOD** | Simple online form. | Online form, and sometimes XML upload. | In addition to online form and XML upload, batch publishing and harvesting are often needed. |
| **ADMINISTRATION** | None or automatic. | Manages accounts and metadata. | |
| **SECURITY** | Varies with specific requirements. | | |
| **DISCOVERY METHOD** | Browsing can be sufficient. | Search or browse. | Search is critical. |
| **RESULT RANKING** | Not important because of the small number of records. | Important. | Very important as a search may return hundreds or more matching records. |
| **PERFORMANCE TUNING** | Not an issue with the small number of records. | Very important when the size of the catalog is big. | |

**Table 6.2**   Geoportal design considerations

## 6.3.1 PERSONAL AND DEPARTMENTAL GEOPORTALS

ArcGIS Services Directory demonstrates the main features provided by personal and departmental geoportals. ArcGIS Services Directory is included with ArcGIS Server. It allows you to browse through the geospatial Web services available from an ArcGIS server, review the service metadata, preview the available services, and obtain information for connecting to the services. A personal or departmental geoportal typically has the following features:

- **Simplified metadata format:** Lengthy metadata can involve too much overhead, and standards are not as important for a portal that is usually used internally for a small audience. When you publish ArcGIS Server Web services, only the service name and a short description are needed. Other service metadata such as spatial extent, spatial reference, map tiling schema (for cached map services), and layer information are populated automatically by ArcGIS Server (figure 6.8).

- **Browse instead of search:** Because a personal portal typically has a small number of records, it usually organizes contents into a hierarchical directory that allows you to browse for needed services. The ArcGIS Services Directory organizes Web services in a hierarchy of folders. By following the links, you can navigate from folders to subfolders to find metadata about services and layers.

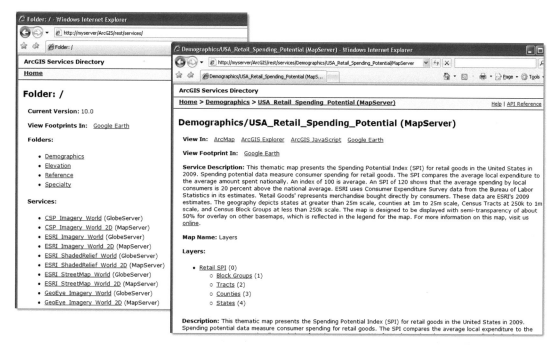

**Figure 6.8** The ArcGIS Services Directory, a personal or departmental geoportal designed to work with ArcGIS Server Web services, allows you to conveniently browse through the services available on ArcGIS Server, examine the services metadata, and preview and experiment with services.

- **Focus on convenience:**

  - Metadata records are automatically published, updated, and deleted as you create, update, and delete Web services in ArcGIS.

  - URL links provided in service metadata pages allow you to preview the service in ArcGIS Desktop, ArcGIS Explorer, an ArcGIS JavaScript map viewer, Google Earth, Google Maps, or Microsoft Bing Maps.

- **Other features:** The ArcGIS Services Directory has built-in security. Secured services are visible only to authorized users, who can see services metadata and preview it.

## 6.3.2 ORGANIZATIONAL AND COMMUNITY GEOPORTALS

ArcGIS.com is an online GIS that includes a geoportal. This geoportal makes it easy for the ArcGIS user community to contribute, find, and use geospatial contents. It represents the characteristics of typical organizational and community geoportals, including the following:

- **Rich content for the community:** ArcGIS.com provides a variety of resource types that work with the ArcGIS suite of products, including the following:

  - ArcGIS Server 2D/3D map services, image services, feature services, ArcIMS services, and Microsoft Bing Maps.

  - ArcGIS Server query, locator, place finder, routing, geometry, and geoprocessing services.

  - Tools, scripts, and add-ins to ArcGIS products. Examples include a Yellow Pages-type search that can find nearby businesses, weather forecasts, and links to Google Street View or Microsoft Bird's Eye View that can be added to ArcGIS Explorer.

  - Layer packages, which are sets of data, plus renderers such as the classification and symbolization scheme. Users of ArcGIS Desktop and ArcGIS Explorer can download and view these files drawn as the map designer saved them.

  - Map packages, which include map documents (MXD) along with associated data.

  - Web mapping application and mobile mapping applications.

- **Simple metadata format for easy publishing:** The ArcGIS.com geoportal (figure 6.9) uses a simplified metadata format so that publishers can quickly contribute contents. For example, when contributing a map service, a publisher only needs to fill in the Web service access URL, title, and tags (see section 6.2.2). The geoportal automatically extracts the spatial reference and other information from the Web service without requiring the publisher to provide them. Publishers can provide additional information, if desired, with the metadata editing tools available in ArcGIS Online.

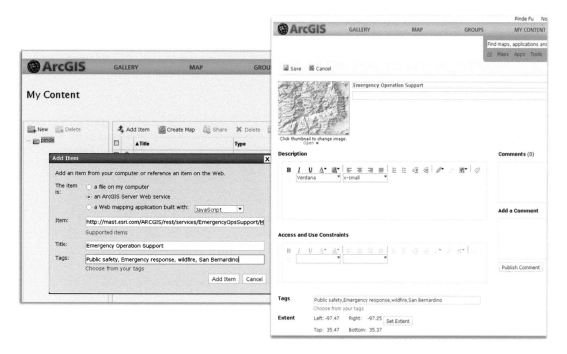

**Figure 6.9** ArcGIS Online uses a simple metadata format. Publishers only need to provide a few pieces of information, at left, to quickly contribute content. Publishers have the choice, at right, to edit the metadata to specify more information.

- **Search for efficient discovery:** Searching and browsing complement each other, suiting different situations. With thousands of metadata records to look through, users won't want to look at every one. Browsing is efficient when the records are already classified, the classification criteria is well designed, and the user knows in advance which category the item he or she is looking for belongs to. If the user doesn't know the classification or has a different view than the cataloger about the classification, the user won't be able to find the item. Searching for an item is typically more efficient than browsing because the records don't have to be precategorized and users don't need to know any classification criteria. The computer scans through all the records, finds the matching ones, and presents the results to the user. ArcGIS.com, expected to hold a large number of records, provides users with a search method for efficient resource discovery.

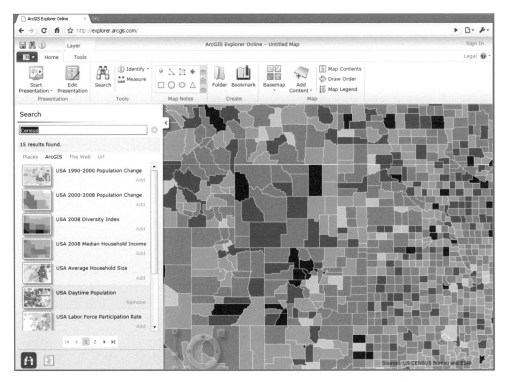

**Figure 6.10**   The search capability of ArcGIS.com is integrated with ArcGIS products. Illustrated here, a user searches for contents and directly adds them via ArcGIS Explorer Online.

Courtesy of U.S. Census Bureau.

- **Seamless integration with ArcGIS products:** ArcGIS.com provides a set of functions that are deeply integrated with ArcGIS to aid user convenience. After discovering relevant contents, you can immediately view the contents and create composite maps by simply clicking the links provided. You can also search for contents in ArcGIS Explorer and directly add contents (figure 6.10) or create composite maps using the default.

- **Other features:** The ArcGIS.com geoportal provides a simple and clear approach to requiring access control. Publishers can join existing groups, create new groups, and grant access to others who want to be in their group. Publishers who contribute resources can control which groups have access to them.

The ArcGIS.com geoportal is an example of a community geoportal that has a strong emphasis on user convenience. It allows members to easily organize, find, and share geographic information and collaborate with other ArcGIS users who share a common interest. The geoportal allows users to jump-start their GIS projects with access to rich ready-to-use contents for use in building applications.

## 6.3.3 REGIONAL, NATIONAL, AND GLOBAL GEOPORTALS

Geospatial One-Stop (GOS) (`http://www.geodata.gov`) is an example of regional, national, and global portals. As a U.S. national-level geoportal, GOS provides a single point of access to hundreds of thousands of metadata records on map services, downloadable datasets, images, map files, and data clearinghouses. As one of twenty-four e-government initiatives sponsored by the federal Office of Management and Budget, GOS goals include the following:

- Developing a portal for seamless access to geospatial information

- Providing standards and models for geospatial data

- Creating an interactive index to geospatial data holdings at the federal and nonfederal level

- Encouraging greater coordination among federal, state, and local agencies on existing and planned geospatial data collecting

GOS, released in 2003, is based on an early version of the ArcGIS Server Geoportal Extension (formerly GIS Portal Toolkit). The metadata records are submitted to the portal by government agencies at all levels (federal, state, local, and tribal), companies, academia, nonprofit organizations, and individuals. Users looking for data can specify a number of criteria to search for relevant resources and browse through featured data and map services on related community pages. The metadata within the portal can then direct users to the actual data sources. Many of the data sources can be directly downloaded and many of the Web services can be viewed with the GOS map viewer. As a major source of geospatial information on the United States, GOS has a large catalog of 430,000 records as of May 2010. GOS represents the typical characteristics of regional, national, and international geoportals, including the following:

- **Emphasis on interoperability and standards:** Considering the variety of software products used within a region, country, or multiple countries, it is important for large geoportals to comply with national, international, and industrial standards. GOS supports standards on several levels: It supports metadata in FGDC and ISO 19915 standards; it provides convenient integration with resources, including OGC WMS, WFS, WCS, KML, and GeoRSS; and users can use a variety of clients to view and use these standard resources (figure 6.11).

- **Search is essential given the large number of records:** Considering the hundreds of thousands of records in GOS, search is a must. The GOS search interface allows users to specify keywords, content type (Web services, online data, offline data, etc.), data category (i.e., themes such as agriculture and farming, biology and ecology, atmospheric and climatic, business and economic, geological, elevation, health, imagery, water, etc.), area of interest, time span, and publisher (figure 6.12). In addition to search, GOS helps to create communities that focus on specific interests, such as fire mapping, homeland security, and hurricanes. Each community maintains a list of featured data and services for users to quickly access.

- **Result ranking:** It is common for a single search in a large geoportal to return many pages of results. Users expect to see the contents they really need within the first page of results, making result ranking an important feature of large geoportals. GOS allows users to sort the results by relevance, title, date, and size of area.

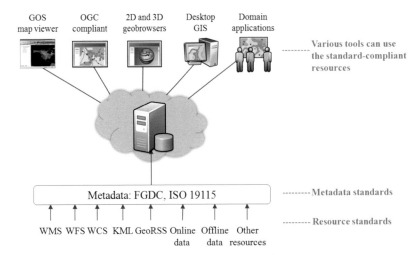

**Figure 6.11** GOS supports metadata in FGDC and ISO standards and provides convenience in publishing OGC standard resources. Users can use a variety of tools rather than being tied to a single tool to use the GIS resources discovered through a GOS portal.

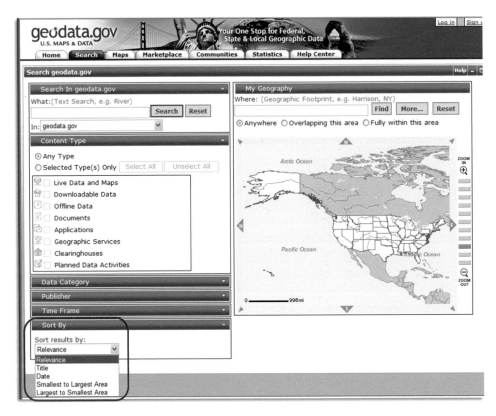

**Figure 6.12** The GOS search interface allows users to specify criteria such as what, where, and when. It also allows users to specify how the search results should be sorted (see red box).

Courtesy of U.S. Geological Survey.

- **Metadata harvesting:** Contributors to a national geoportal typically include government and academic organizations that already have their own metadata catalogs. For example, GOS has thousands of registered publishers, and a few hundred of them have their own metadata catalogs or geoportals (figure 6.13). Some publishers, such as the U.S. Census Bureau, National Oceanic and Atmospheric Administration (NOAA), and National Park Service (NPS), have tens of thousands of metadata records each in their catalogs. It would be a tiring and tedious job for each publisher to have to publish their own records, one by one, to the national geoportal, let alone to continuously update each record. GOS resolved this problem by using a harvester (see section 6.2.3). GOS allows publishers to register their metadata catalog, and then GOS regularly and automatically connects to these catalogs, fetches the metadata XMLs, and publishes the metadata in the GOS central catalog.

- **Marketplace:** GOS provides a metadata marketplace to promote data partnerships. Organizations post their planned geospatial data acquisition activities to Marketplace. This enables other organizations to identify the collaboration opportunities for future data acquisition, allows them to leverage resources, maximizes geospatial investments, and reduces redundancies in data acquisition.

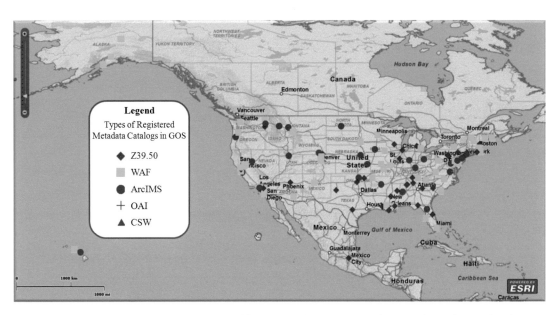

**Figure 6.13**   The organizations that are registered for metadata harvesting and the type of metadata catalog they hold are displayed on a map. GOS periodically harvests metadata from these catalogs and publishes it in the GOS central catalog.

Map data © AND Automotive Navigation Data; courtesy of Tele Atlas North America, Inc.

## 6.4 CHALLENGES AND PROSPECTS

Something as diffuse as NSDI will never be seen as a success by everyone, but it has been a catalyst for many positive developments (Longley et al. 2005). As an important component of NSDI, geoportals have produced invaluable benefits. Geoportals facilitate the sharing of information, eliminate duplicate data collecting, and support more rational decision making. For example, during Hurricanes Katrina and Rita in the United States in 2005, data providers quickly established virtual communities in GOS and contributed critical information, including data, map services, and online map viewer applications. This information was used to aid federal, state, and local agencies in emergency response activities. In China, the Geoportal of Earth System Science Data Sharing Network supported the design and construction of the Qinghai-Tibet Railway, emergency response for the 2008 Wenchuan earthquake, and environmental protection for the 2008 Beijing Olympic Games.

In recent years, geoportal technologies have advanced remarkably, in large part as a consequence of the work and demand of the government, academic, and commercial sectors. Handling metadata complexity, supporting semantic search, protecting copyright, and conducting broader data mining are among the challenges that lie ahead. As the volume of geospatial content increases with the popularity of GPS, geoportals will become even more important. The potential of geoportals to expand the use of geospatial information will be realized more quickly with stronger governmental support and broader public participation.

### 6.4.1 RESEARCH AREAS

The further development of geoportals faces a number of challenges. Some important research areas to facilitate broader use of geoportals include the following sections:

#### HANDLING THE COMPLEXITY OF METADATA STANDARDS

National and international metadata standards have good reason to be complex, since they are generally designed to describe all the properties of geospatial resources. This complexity, however, has made metadata itself a bottleneck for sharing information. It is not uncommon to hear data owners say, "Metadata is too complex," and "I don't have the time and funds to create metadata." For some publishers, the complexity of metadata has been an obstacle to publishing their data quickly and sharing that information. The decision of whether to comply with metadata standards remains a difficult choice for many data owners and geoportal designers.

The solution lies on many fronts. Software vendors can provide enhanced tools that simplify, or even automate the creation of standard metadata as much as possible. The focus of geoportals needs to shift from metadata to "findability"—that is, it is more important for a resource to be easily discovered in a geoportal than it is to have verbose metadata that documents it. This trend is observed in the latest versions of ArcGIS Server and ArcGIS.com. The ArcGIS Server Geoportal Extension can transform metadata into HTML and GeoRSS formats, which can increase the exposure of metadata for easier discovery and broader accessibility. ArcGIS Services Directory and services metadata are HTML pages that can be directly crawled and discovered by general-purpose Web search engines. Emerging search services (see chapter 3) can directly index the geospatial data, including the attribute tables, of an enterprise, making the data easier to discover. ArcGIS.com uses simple metadata tags (see section 6.2.2), thus lowering the barrier to contributed geospatial contents.

## SUPPORTING SEMANTIC SEARCH

**Semantic search refers to a search based on the meaning of keywords rather than the spelling of keywords.** Today's geoportals typically are unable to account for the meaning of keywords, the context in which keywords are used, and the ambiguities of natural language. A strict keyword search generally will not satisfy the needs of a thorough, intelligent search. For example, if you were to search for the keyword "river," you might also want streams, channels, lakes, or other features related to hydrology but not think of the words "river city bank." Searches that rely on a matched string may miss records that you really want while including records that you do not want.

Progress has been made toward semantic search (Yang et al. 2008; Arpinar et al. 2006; Lutz et al. 2004). Geoportal products such as ArcGIS Server Geoportal Extension support certain levels of semantic search by incorporating OWL (Web Ontology Language). **Ontology is a formal representation of a set of concepts within a domain and the relationships among these concepts. OWL is a family of languages for representing ontologies.** For example, GEMET (General Multilingual Environmental Thesaurus) was developed by the European Topic Center on Catalog

```
<owl:Class rdf:about="http://www.eionet.europa.eu/gemet/concept/7244">
    <label xml:lang="de">FluÃŸ</label>
    <label xml:lang="no">elv</label>
    <label xml:lang="it">fiume</label>
    <label xml:lang="fr">fleuve</label>
    <label xml:lang="hu">folyÃ³</label>
    <label xml:lang="eu">ibai</label>
    <label xml:lang="fi">joki</label>
    <label xml:lang="et">jÃµgi</label>
    <label xml:lang="sl">reka</label>
    <label xml:lang="sk">rieka</label>
    <label xml:lang="pt">rios</label>
    <label xml:lang="en">river</label>
    <label xml:lang="en-us">river</label>
    <label xml:lang="nl">rivier</label>
    <label xml:lang="pl">rzeka</label>
    <label xml:lang="es">rÃos</label>
    <label xml:lang="da">Ã¥</label>
    <label xml:lang="cs">Å™eka</label>
    <label xml:lang="el">Ï€Î¿Ï„Î±Î¼ÏŒÏ‚</label>
    <label xml:lang="bg">Ð ÐµÐºÐ°</label>
    <label xml:lang="ru">Ñ€ÐµÐºÐ°</label>
    <seeAlso rdf:resource="http://www.eionet.europa.eu/gemet/concept/10161"/>
    <seeAlso rdf:resource="http://www.eionet.europa.eu/gemet/concept/1288"/>
    <subClassOf rdf:resource="http://www.eionet.europa.eu/gemet/concept/9161"/>
    <subClassOf rdf:resource="http://www.eionet.europa.eu/gemet/group/4125"/>
    <subClassOf rdf:resource="http://www.eionet.europa.eu/gemet/theme/23"/>
    <subClassOf rdf:resource="http://www.eionet.europa.eu/gemet/theme/40"/>
</owl:Class>
```

**Figure 6.14** A small section of GEMET OWL depicts terms related to the word "river."

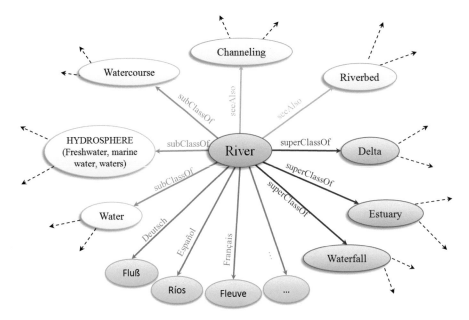

**Figure 6.15**   A conceptual representation of GEMET OWL depicts the terms associated with "river."

of Data Sources under contract to the European Environment Agency. Its OWL includes more than 6,000 terms and defines the relationships among these terms (figure 6.14). OWL can also be presented in diagrams (figure 6.15). When you search for a keyword using the ArcGIS Server Geoportal Extension, the extension searches OWL first, finds related terms, such as synonyms or narrow terms, and then searches the geoportal catalog for records matching these terms rather than just the term you requested. Using OWL can provide a more comprehensive search result. Theoretically, this is still far from a fully realized semantic search, which remains a frontier of geoportal research.

## PROTECTING COPYRIGHT

Digital resource owners share a common concern over protecting copyright. Once the data is given to a user, the owner pretty much loses control of the data. The fear of losing control over the intellectual property rights for their geospatial resources makes many organizations hesitate in participating in geoportals. They hope for effective ways to ensure that their data is only used by authorized users, only for the period authorized, and only for certain types of applications. A Web-service-based information-sharing mechanism can solve these problems to some extent. Other solutions (figure 6.16) include DRM (digital rights management)-related mechanisms. Standards bodies such as OGC have defined specifications related to the GeoDRM Reference Model that can specify payment-based information sharing and data lock mechanisms when the data is used illegally or after the expiration date. However, integrating data copyright protection specifications within the information-sharing workflow of geoportals remains a challenge.

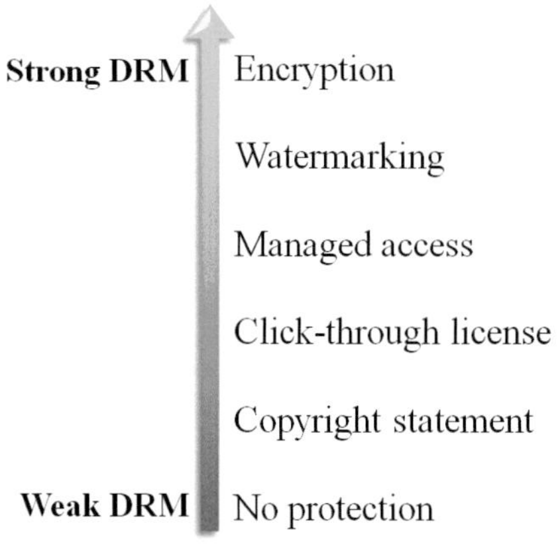

**Strong DRM**
Encryption

Watermarking

Managed access

Click-through license

Copyright statement

**Weak DRM**
No protection

**Figure 6.16** Digital intellectual property can be managed and protected in numerous ways.

## DATA MINING

The data shared in today's geoportals is mostly data that has been collected by government agencies, universities, and professional private-sector organizations. Such authoritative data is extremely valuable. But just as valuable for certain types of applications is the massive amount of geospatial information that is being contributed by casual Web users. Such data is often unstructured, dynamic, and usually lacking the quality of authoritative data yet still useful. How to extract information from the massive Web, and then verify its authenticity, determine its spatial and temporal accuracy, and incorporate it into geoportals is a field that demands research attention.

## 6.4.2 LOOKING FORWARD

The volume of geospatial information in recent years has been boosted by advances in GPS, remote sensing, and other survey technologies that have made it easier and faster for government, the private sector, and even individuals to collect and disseminate geospatial data. **The more information that is available, the harder it is to locate a particular piece of it, and the stronger the need for geoportals.**

The formats used to share geospatial information have also advanced and diversified in recent years. **Web services have many advantages as a means of sharing geospatial information.** Data providers can control access to the information, and thus can better protect intellectual property rights. Users can reduce the time and expense in collecting data and acquiring computer systems, gaining flexibility in quickly building applications on top of Web services (see chapters 3 and 7). **Geoportals should effectively facilitate the discovery and use of geospatial Web services.**

Tim Berners-Lee, the father of the Web, envisions that the next generation of the Web will be the Semantic Web (see section 10.2.2). **In the Semantic Web, the meaning (semantics) of information is defined, making it possible for machines to process it** (Berners-Lee, Handler, and Lassila 2001; W3C 2001). While the full realization of the Semantic Web is still a long way off, its implementation is expected to reduce the vagueness, uncertainty, and inconsistency of natural language, making the discovery of geospatial information more intelligent and accurate.

**The sharing of information is not only a technical issue, but a social issue as well.** Information is power. Sharing information with the right parties multiplies the power of society overall. Some countries restrict the sharing of information because of national security concerns. However, with the technological advances in GPS, even civilian GPS units on the market can reach an accuracy of 10 meters, which is more than sufficient to produce maps on a scale of 1:5,000. And with Google Earth and Microsoft Bing Maps serving submeter resolution imagery for many regions of the world, limiting public access to such similar geospatial data has little meaning in terms of national security, but rather serves to limit the applications of GIS to serve society (Ye 2004). Still, some organizations are not motivated to share data because of the fear that they will lose control over it. This is one of the reasons that governments need to promote the sharing of information through policy, law, and regulation (Onsrud 1995). Under the U.S. Freedom of Information Act, signed into law by President Johnson in 1966, federal government agencies are duty bound to provide government records and data, which includes geospatial data, to the public free of charge or at minimum cost, if the data does not invade personal privacy or national security. Today, more than 100 countries have developed similar legislation or policies that endorse SDI. Such legislation and policy will serve as a strong basis for future geoportal research and development. In the future, the more intelligent geoportals will support easier, faster, and more accurate discovery of geospatial resources and foster more applications of GIS.

## Study questions

1. What are geoportals? Why do we need them?

2. What is NSDI? What is the role of geoportals in NSDI?

3. In a typical geoportal, there are three types of players. What are their roles? What functions does each player need to have?

4. Define geospatial metadata, explain its uses in the context of geoportals, and list several metadata standards.

5. Describe the need for metadata 2.0 and explain its characteristics and uses.

6. Why is metadata harvesting or synchronization needed? How does it work?

7. What are the characteristics or design considerations for a national geoportal?

8. Discuss the challenges that face today's geoportal development.

9. Define semantic search, ontology, OWL, and the Semantic Web.

10. Outline the prospects for geoportals.

## References

Arpinar, Budak, Amit Sheth, Cartic Ramakrishnan, E. Lynn Usery, Molly Azami, and Mei-Po Kwan. 2006. Geospatial ontology development and semantic analytics. *Transactions in GIS* 10 (4): 551–75.

Beaumont, Peter, Paul A. Longley, and David J. Maguire. 2005. Geographic information portals: A UK perspective. *Computers, Environment and Urban Systems* 29:49–69.

Bernard, Lars, Ioannis Kanellopoulos, Alessandro Annoni, and Paul Smits. 2005. The European geoportal: One step towards the establishment of a European Spatial Data Infrastructure. *Computers, Environment and Urban Systems* 29:15–31.

Berners-Lee, Tim, James Handler, and Ora Lassila. 2001. The Semantic Web. *Scientific American.* `http://www.sciam.com/article.cfm?articleID=00048144-10D2-1C70-84A9809EC588EF21` (accessed November 11, 2009).

FGDC. 2000. *Content standard for digital geospatial metadata workbook.* `http://www.fgdc.gov/metadata/documents/workbook_0501_bmk.pdf` (accessed September 22, 2009).

———. 2006. Geospatial metadata. `http://www.fgdc.gov/metadata` (accessed September 20, 2009).

Goodchild, Michael F. 2004. The Alexandria Digital Library Project: Review, assessment, and prospects. *D-Lib Magazine* 10 (5). `http://www.dlib.org/dlib/may04/goodchild/05goodchild.html` (accessed September 2, 2009).

———. 2008. Spatial accuracy 2.0. Proceedings of the 8th international symposium on spatial accuracy assessment in natural resources and environmental sciences. Shanghai, June 25–27. `http://www.geog.ucsb.edu/~good/papers/453.pdf` (accessed February 4, 2010).

Goodchild, Michael F., and Jianyu Zhou. 2003. Finding geographic information: Collection-level metadata. *GeoInformatica* 7 (2): 95–112.

Goodchild, Michael F., Pinde Fu, and P. M. Rich. 2007. Geographic information sharing: The case of the Geospatial One-Stop portal. *Annals of Association of American Geographers* 97 (2): 249–65.

Hogeweg, Marten. 2009a. Spatial data infrastructures. ESRI International User Conference, San Diego, Calif., July 13–17.

———. 2009b. SDI for everyone. `http://martenhogeweg.blogspot.com/2009/10/sdi-for-everyone.html` (accessed February 4, 2010).

Li, Fangfang. 2006. Spatial information sharing and services and enlightenment in developed countries. *Land Resources Informatization*, no. 4:45–8.

Longley, Paul A., Michael F. Goodchild, David J. Maguire, and David W. Rhind. 2005. *Geographic information systems and science.* 2nd ed. San Francisco: John Wiley & Sons.

Lutz, Michael, Udo Einspanier, Eva Klien, and Sebastian Hübner. 2004. An architecture for ontology-based discovery of geographic information. Proceedings of the 7th AGILE Conference on Geographic Information Science, Herkalion, Greece, April 29–May 1.

Maguire, David J., and Paul A. Longley. 2005. The emergence of geoportals and their role in spatial data infrastructures. *Computers, Environment and Urban Systems* 29:3–14.

Mapping Science Committee, National Research Council. 1993. *Toward a coordinated spatial data infrastructure for the nation.* Washington, D.C.: National Academies Press.

Masser, Ian. 2007. *Building European spatial data infrastructures.* Redlands, Calif.: ESRI Press.

National Research Council. 1993. *Toward a coordinated spatial data infrastructure of the nation.* Washington, D.C.: National Academies Press.

Nebert, Douglas D., ed. 2004. *Developing spatial data infrastructures: The SDI cookbook.* `http://www.gsdi.org/docs2004/Cookbook/cookbookV2.0.pdf` (accessed September 16, 2009).

Onsrud, Harlan J. 1995. The role of law in impeding and facilitating the sharing of geographic information. In *Sharing geographic information,* ed. H. J. Onsrud and G. Rushton, 292–306. Rutgers, N.J.: CUPR Press.

Smith, Terence R., and James Frew. 1995. Alexandria Digital Library. *Communications of the ACM Archive* 38 (4).

Sun, Jiulin, and Huizhong Shi. 2002. *Science data management and sharing.* Beijing: China Science and Technology Press.

Tait, Michael G. 2005. Implementing geoportals: Applications of distributed GIS. *Computers, Environment and Urban Systems* 29:33–47.

Tang, Winnie, and Jan Selwood. 2005. *Spatial portals: Gateways to geographic information.* Redlands, Calif.: ESRI Press.

The White House. 1994. Executive Order 12906: Coordinating geographic data acquisition and access: The National Spatial Data Infrastructure. `http://govinfo.library.unt.edu/npr/library/direct/orders/20fa.html` (accessed November 11, 2009).

Van Oosterom, Peter, and Sisi Zlatanova. 2008. *Creating spatial information infrastructures: Towards the spatial Semantic Web.* Boca Raton, Fla.: CRC Press, Taylor & Francis Group.

W3C. 2001. W3C Semantic Web frequently asked questions. `http://www.w3.org/2001/sw/SW-FAQ` (accessed November 11, 2009).

Wagner, Roland M. 2005. Open Geospatial Consortium GeoDRM. 2nd ODRL workshop, Lisbon, Portugal, July 7–8. `http://odrl.net/workshop2005/program/prez-wagner.ppt` (accessed September 25, 2009).

Wang, Juanle. 2005. The study of the metadata key issues in Scientific Data Exchange Center. PhD diss., Chinese Sciences Academy.

Yang, Chaowei, Wenwen Li, Jibo Xie, and Bin Zhou. 2008. Distributed geospatial information processing: Sharing distributed geospatial resources to support Digital Earth. *International Journal of Digital Earth* 1 (3): 259–78.

Ye, Jia'An. 2004. Proposal for Geographic Information Data Special Administrative Region: How to have GIS better serve social and economic development. *Earth Information Science* 6 (3): 1–3.

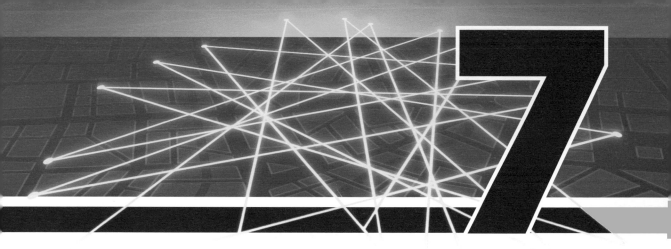

# NSDI IN THE WEB 2.0 ERA

National Spatial Data Infrastructure (NSDI), or National Spatial Information Infrastructure (NSII), encompasses the policies and technology used within a nation to produce, manage, and share geospatial information. NSDI promotes collaboration at the local, regional, and national levels. The concept originated in the United States in the early 1990s and has been adopted by most countries over the last two decades. NSDI is a massive and diffuse subject. This chapter focuses on the technical side of information sharing and collaboration, which is one of the most fundamental elements of NSDI.

The first decade of building NSDI was primarily government-centric and relied on data duplication to share geospatial information. Agencies largely depended on FTP sites to collaborate on data. The time delays and high costs of receiving data through the data duplication method have made it a barrier to potential uses of geospatial information. Web 2.0 provides new principles and technologies that facilitate information sharing and collaboration, including the bottom-up flow of information, open and interoperable Web services, Web-oriented architecture (WOA), mashups, and cloud computing. The new generation of NSDI—NSDI 2.0—should pursue these advantages for more efficient ways to share geospatial information.

The feasibility and benefits of building NSDI 2.0 by expanding the use of Web 2.0 technologies are reflected in a number of initiatives around the world and demonstrated by the success of numerous applications. Web services, cloud computing, mashups, geobrowsers, and other Web 2.0 technologies vastly increase the flexibility of sharing information and encourage user-generated content (UGC), and these technical advances should work in favor of NSDI. To be most effective,

NSDI 2.0 would do well to take a decentralized approach, with the private sector and individual citizens playing important roles. It should have a distributed architecture just like the Web that facilitates data production and maintenance. Geospatial resources (data, maps, and tools) should be shared as open, standardized, and easy-to-consume Web services as much as possible to facilitate being easily reused and remixed via mashups, geobrowsers, and other applications. Web services should be made available to other organizations and to citizens when appropriate, thus maximizing the value of gespatial resources.

This chapter has four sections. Section 7.1 presents the advantages of the Web-2.0-based approach over traditional data duplication. Section 7.2 discusses data collecting, by organizations and by private citizens, and data dissemination via Web services and cloud computing. It also points out issues with current Web services. Section 7.3 explains the integration of Web services using mashup-style programming and geobrowsers, and section 7.4 highlights the challenges to building NSDI 2.0 and the outlook for the future.

## 7.1 FROM DATA DUPLICATION TO WEB SERVICES

Sharing information is an important component of NSDI. The approach used to share information is advancing from data duplication to being Web service-centric.

### 7.1.1 SHARING INFORMATION IN NSDI

The term SDI was originally highlighted in "Toward a coordinated spatial data infrastructure of the nation" by the U.S. National Research Council (1993) and in President Clinton's Executive Order 12906, "Establishing NSDI in the United States" (The White House 1994), in which the **NSDI is defined as "the technology, policies, standards, and human resources necessary to acquire,**

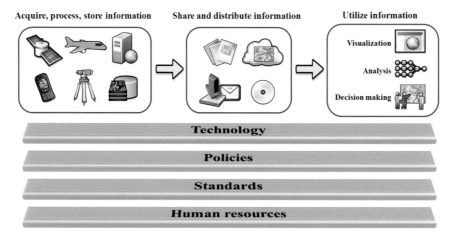

**Figure 7.1**  Overview shows how the components of NSDI work together to promote the sharing of geospatial information.

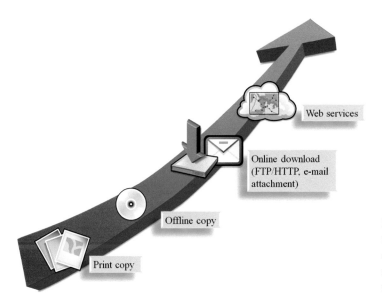

**Figure 7.2** Geospatial information can be shared by a variety of means. The use of Web services is an increasingly popular way to share information.

**process, store, share, distribute, and improve utilization of geospatial data"** (figure 7.1). From the perspective of the information sciences, data is a source and a carrier of information. Therefore, NSDI and NSII are often interchangeable (Wu and Tong 2008).

One of the most important functions of NSDI is to promote the sharing of geospatial information throughout government at all levels, the private and nonprofit sectors, and the academic community. By nature, geospatial information is distributed, heterogeneous, and expensive to collect. Rarely can all the necessary information be found in a single database in a single data format. GIS users often need to rely on outside sources for GIS data. Hence, **"collect once and reuse many times" is one of the main philosophies behind NSDI.** The traditional approach to sharing and collaborating on geospatial information has been to use data duplication and printed map copies. While these approaches are still needed in many situations, they have their limitations, and NSDI should move toward a Web-service-centered approach (figure 7.2).

## 7.1.2 DATA-DUPLICATION-BASED NSDI 1.0

Traditionally, organizations that needed to work with GIS had to acquire GIS hardware and software, hire professional GIS staff, and obtain duplicate copies of geospatial data before they could benefit from GIS technology. This is still the case in many situations. In a typical scenario (figure 7.3), the data provider needs to export the data into files of a certain format, transfer it onto magnetic tapes or optical disks (e.g., CD-ROM or DVD), and deliver the data to other agencies via postal service or electronic transfer (e.g., FTP, HTTP, or e-mail). Once an organization receives the data, it has to import it into a local database, and then build its applications.

Data duplication is still needed in certain situations and can be used successfully, but it has the following disadvantages:

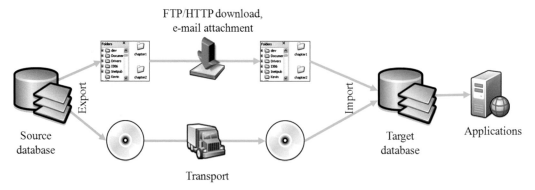

**Figure 7.3** Data duplication, which involves exporting, transporting, and importing the data, is often not an efficient way to share information.

- **Delays:** The export, transport/download, and import process usually takes time, typically from hours to weeks. The data can become outdated during that period. Delays in obtaining data are not desirable for many of today's time-sensitive applications such as emergency response, where the situation can change by the minute (Zhao, Yu, and Di 2007). In some cases, such as in large collections of imagery, which are measured in terabytes, duplicating and transferring the entire dataset is not practical.

- **High cost for the recipients:** Receiving organizations must have the right hardware, compatible software, and knowledgeable staff to be able to import the data, tune the database, and work with the data. This is costly in terms of labor and infrastructure and is often cost-prohibitive for nonprofessional and small organizations that need GIS.

- **Copyright concerns:** The owner of the source data loses control of that data once it is duplicated by others. Organizations that receive the data may use it in ways that were not intended by the provider, such as using it in an unauthorized scope or copying it to others without permission. Because of the fear of losing intellectual property, organizations that own data may not be motivated to share it.

The disadvantages of having to rely on data duplication have hindered the broad and efficient use of GIS. It has proved to be an obstacle to emergency management teams, which need real-time data from multiple agencies in real time to make immediate decisions; to restaurant owners, who usually have no idea what GIS is but want to choose the most profitable locations for new sites; and to citizens who want to quickly and easily assemble different sources of information about their home or points of interest. For society as a whole, data duplication often results in duplicated efforts and wasted funding. This approach to implementing NSDI needs to be reengineered (Binns 2007).

## 7.1.3 WEB-SERVICES-BASED NSDI 2.0

The adoption of Web services and service-oriented architecture (SOA), along with the more recent WOA, cloud computing, mashups, geobrowsers, and technologies that encourage UGC, are providing new ways to share information and foster collaboration in the Web 2.0 era. Geobrowsers, Web services, and mashups are introduced in chapters 2, 3 and 4. Cloud computing will be discussed in chapter 10. This section discusses UGC, SOA, and WOA.

Human beings can be thought of as distributed sensors that collect information on what is going on right now in every corner of the world, like backup file disks that remember what it is like in all the places they have been. Before Web 2.0, there were no efficient ways for private citizens to disseminate this information. Web 2.0 facilitates user participation, social networking, and crowdsourcing. This has resulted in massive amounts of UGC and the phenomenon of a bottom-up information flow over the Web. UGC in the geospatial context is termed VGI (volunteered geographic information, see section 10.1). The public's immense capability to contribute VGI is exhibited in a number of Web 2.0 sites such as Wikimapia and OpenStreetMap. While NSDI 1.0 has been a government-centric undertaking with government being the main data collector and provider, VGI, and the broad participation of citizens and businesses in the private sector, has diminished government's central role in NSDI 2.0.

SOA is an architectural style that involves the use and reuse of software components. Professionals use software components to build information systems, very much like kids use building blocks to create their forts and castles. The different types of software development components have evolved over the years (figure 7.4). In the early years of programming, functions and subroutines were the building blocks of software development. In the early 1990s, developers started adopting object-oriented programming, where objects were the building blocks. In the late 1990s, a new kind of "block" was introduced via service-oriented architecture (Schulte and Natis 1996).

SOA has various definitions. Basically, **SOA is a software architecture style in which business functionalities are packaged as a collection of loosely coupled and distributed services, or units, that can be flexibly combined and reused to produce new software applications.** The reusable components in SOA should be aligned with business processes. Their granularity, or size in terms of functionality of a service, is typically larger than components such as subroutines and objects since a service is generally made up of multiple objects and subroutines. Early SOA systems were implemented using CORBA (common object request broker architecture) and DCOM (Distributed

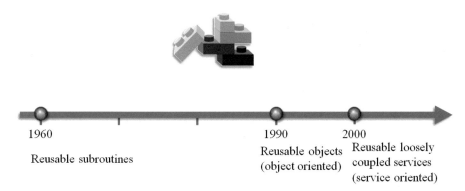

**Figure 7.4**  As information technology has evolved, programming components have also advanced, although the newer component technologies do not replace the older ones.

Component Object Model), but those types of architectures did not fully deliver the flexible and easy integration that SOA promises.

The adoption of Web services presented an ideal way to implement SOA because Web services can be easier to deploy and integrate in a distributed network environment. SOAP-based Web services still could not take full advantage of the Web, however, because of their inability to use HTTP cache and complexities such as those introduced by the SOAP envelope that wraps the request and response XMLs (see chapter 3). As an important complement, RESTful Web services are primarily based on URLs and can take full advantage of the HTTP caches. Without the SOAP envelope, these Web services can be easily invoked and integrated in mashups. RESTful Web services, cloud computing, and lightweight mashups demonstrate how the SOA tenets of flexibility, reusability, and reduced complexity can be achieved over the Web by relatively simple means. This new usability led to the birth of WOA. **WOA represents a specialization of SOA that uses RESTful Web services and lightweight mashups** (Thies and Vossen 2008). WOA can create supple and resilient software that is less difficult and less expensive to build than traditional SOA implementations (Hinchcliffe 2008; Gall 2008; Kralidis 2007). The SOA approach, including WOA, has several big advantages over the data duplication approach:

- **Cost-effectiveness for client organizations:** Client organizations can use many services for free or pay by subscription or for the actual amount of use, which is typically cheaper than acquiring and installing GIS hardware, software, and data on a local basis.

- **Greater data currency:** When the source database is updated, updates are immediately reflected in Web services and available to all client applications.

- **Greater flexibility and agility:** WOA can take advantage of the Web, deliver the flexibility and agility that SOA has aimed for, and support dynamic integration of Web services through the use of mashups or geobrowsers. This capability substantially reduces the difficulty in reusing geospatial data and functions.

- **Easier interoperability:** Data format is no longer a barrier. With industry-standard Web services, geospatial information can be directly absorbed into various applications without going through the export/transport/import process.

- **Better protected intellectual property:** Maintaining property rights is easier to address. A provider can share his or her service for free or for profit, charging users by an annual subscription fee or based on a per-request price model.

Because of these advantages, SOA can be the technical architecture for developing the second generation of NSDI, or NSDI 2.0. For example, many countries sponsor SOA-based NSDI research and are developing SDI based on Web services. The impetus to build NSDI based on WOA and Web services is advocated in a number of proposals and articles.

Jeff Harrison of the Carbon Project and others in NSDI 2.0: Powering Our National Economy, Renewing Our Infrastructure, Protecting Our Environment (2009) stress that

*"The technical philosophy of NSDI 2.0 is that geospatial data and environmental information should be maintained locally, closest to source, and then shared with the nation through online services."*

Lisa Campbell of Autodesk and others in A Proposal for Reinvigorating the American Economy through Investment in the U.S. National Spatial Data Infrastructure (2009) emphasize the importance of SOA to NSDI:

*"An NSDI integrates information from many sources and authors using standardized protocols (particularly an Open Geospatial Consortium–compliant service-oriented architecture) so that geospatial data and spatially enabled business data can be harmonized and integrated into a common framework capable of supporting multiple missions at all levels of government and a wide variety of private business initiatives."*

Jack Dangermond of ESRI and Anne Hale Miglarese of Booz Allen Hamilton in An Investment in Geospatial Information Infrastructure: Building a National GIS (2009) talk about the need for a modern NSDI and the supporting technologies:

*"Existing modern GIS server technology, together with open standards and services-oriented architecture, can provide enabling components for a national GIS immediately."*

Dangermond expounds on moving beyond file sharing to a network of geospatial services in an interview in *Government Computer News* (Kash 2009):

*"The architectural answer for an integrated geospatial framework is not to put all the data into one big database. It will involve creating a network of distributed geospatial services that can be dynamically integrated using open standards and free APIs that can visualize, query, and support advanced applications on the Web."*

As a brief summary, **NSDI 2.0 will become decentralized as the private sector and individual citizens play increasingly important roles in data collecting and sharing services. It should be based on industry-standard Web services and SOA principles to allow easy and flexible information reuse and remixing.** This philosophy can benefit both corporations and individual citizens, who will be able to use mashups and geobrowsers to assemble Web services for their own applications. This vision, when aided by effective policies and standards, will enable organizations and citizens to share and collaborate more flexibly and allow the flourishing of federated information systems.

## 7.2 SHARING WEB SERVICES

Web 2.0 technology allows everyone to contribute geospatial information. Governments and big companies can contribute, and so can individual citizens. Geospatial information can be shared via Web services, via an organization's own infrastructure, or another organization's infrastructure or cloud.

## 7.2.1 CONTRIBUTING GEOSPATIAL DATA

NSDI needs to involve as many organizations and individuals as possible to facilitate expanded sharing and collaboration. NSDI 1.0 has been mostly viewed as a government undertaking. In NSDI 2.0, the private sector and individual citizens are playing increasingly important roles.

Nonetheless, **government plays a crucial role in NSDI, especially in countries where survey- ing and mapping are strongly controlled by the government.** Various levels and branches of gov- ernment hold many types of authoritative information for their jurisdictions. For example, the U.S. Department of Transportation holds real-time highway traffic information; the Centers for Disease Control and Prevention has the latest information on H1N1 cases and their distribution; the Department of Homeland Security monitors where terrorists may strike next; the National Weather Service has current and forecast weather information; USGS records real-time earthquake and other natural disaster information; USFS gauges the real-time extent of wildfires via remote sensing; the U.S. Census Bureau maintains massive national demographic information; and the Bureau of Labor Statistics keeps track of American consumers' buying habits. Local governments

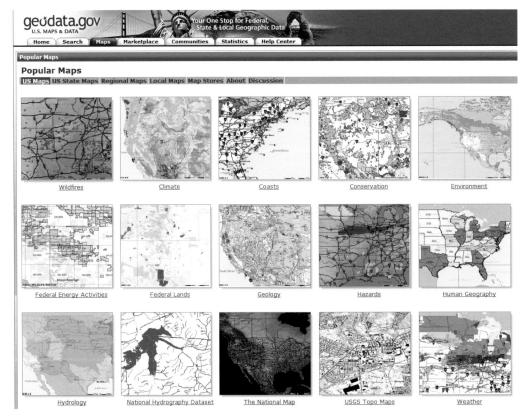

**Figure 7.5** Many government agencies share their authoritative data via Web services. Many popular services are featured on geodata.gov.

Courtesy of U.S. Geological Survey, USFS, NWS, EPA, NOAA, BLM, and U.S. Census Bureau.

often have the most current and accurate data available on local land records, property taxation rolls, and planned zoning. Government agencies should strive to share the information they collect with selected agencies or the public, and take advantage of Web services (figure 7.5).

**The private sector has a large variety of data and expertise to share as well.** For instance, companies such as Tele Atlas, NAVTEQ, and others have frequently updated high-resolution street basemaps and ground imagery; InfoUSA has a database of more than 10 million U.S. businesses, including names and locations; Mediamark Research & Intelligence collects information about millions of adult consumers' media choices, product usage, demographics, lifestyles, and attitudes; the National Research Bureau has a comprehensive source of information about shopping centers in the United States; CMC International has current and extensive U.S. traffic counts at points ranging from freeways to rural roads; and the Inrix Dust Network tracks some 500,000 vehicles on U.S. highways to collect and provide real-time data on traffic congestion.

Sharing information also fosters value-added services. When governments make their data open and freely available, other organizations can innovate from it to create new services (figure 7.6). For

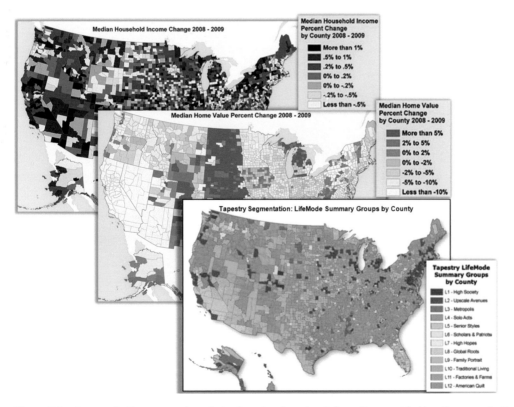

**Figure 7.6** An organization can produce derived services based on information provided by other organizations. Here, ESRI applied algorithms to census and other data sources to produce Tapestry Segmentation services and a series of forecasted demographic map services.

Courtesy of U.S. Census Bureau.

example, ESRI uses census demographic data that is freely available, and then applies economic and statistics models to create Web services on tapestry segmentation, median household income, and median home price percentage change. These Web services are available at ArcGIS Online and ArcGIS Business Analyst Online. Academic organizations collect various types of data for research projects, and often have specialized analysis models that can be used in many practical applications. Thus, not only an organization's data, but its expertise can be desirable commodities that other organizations can use.

**VGI provided by citizens is playing an increasingly prominent role in NSDI.** Citizens have traditionally been neglected in the data-collecting stage of NSDI, but this is no longer the case in the Web 2.0 era. Web 2.0 features a bottom-up information flow as a result of broadened user participation and increased amounts of UGC. Wikimapia, for example, has 12 million places marked and described by citizens. Web albums such as Flickr, Picasa, and Panoramio have millions of georeferenced photos that can help others to visualize places, scenery, and events worldwide. OpenStreetMap is building a patchwork of public-domain street maps contributed by individuals uploading their GPS data. Many social network services allow you to post georeferenced messages. For example, citizens can report nonemergency issues such as potholes and graffiti by marking them on a map and describing them via a Web application (figure 7.7). Section 10.1 summarizes a list of Web sites that encourage VGI.

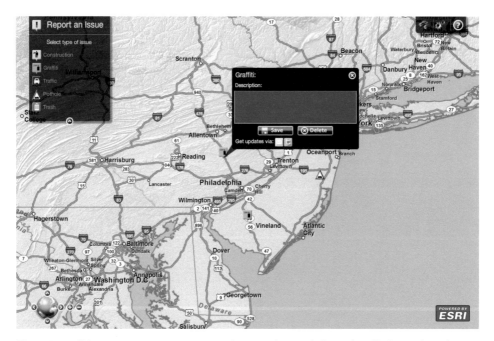

**Figure 7.7**  Citizens can report nonemergency issues such as potholes and graffiti by marking them on a map and describing them. The data collected can be stored in a database via the ArcGIS Server feature edit service and then immediately disseminated via data or map services.

Map data © AND Automotive Navigation Data; courtesy of Tele Atlas North America, Inc.

VGI can make unique contributions to NSDI, although mechanisms for ensuring VGI quality and protecting personal privacy are still needed. The VGI collected by these Web sites can be distributed via Web services for other organizations to use to build new applications.

## 7.2.2 FRAMEWORK SERVICES AND SPECIALIZED SERVICES

According to WOA/SOA principles, Web services should be aligned with business functions. By adopting this principle, NSDI 2.0 will align the Web services produced by an organization with the collective needs of the community. Under the ambitious goals of NSDI, sharing information should be pursued to the greatest degree possible. A Web service should be used not only by the organization that creates it, but by external organizations as well. An organization should determine what Web services to create by analyzing not only its own internal needs, but the needs of external entities as well. Table 7.1 illustrates a way to assess these needs. By creating such a table, a city can list Web services needed by different city departments and those needed by external organizations. These Web services can be categorized as universal use, broad use, and departmental use. Web services A and B in the table are needed by all parties, representing universal use. Web services C, D, and E are needed by many parties, and so are for broad use. Web services F, G, and H are needed by only select parties, or for departmental use. This type of needs assessment can help an organization analyze what services to create, in what priority, by which department, and for which stakeholders, and even how to finance the work.

| DEPARTMENTS NEEDING THE SERVICE | WEB SERVICES | | | | | | | |
|---|---|---|---|---|---|---|---|---|
| | A | B | C | D | E | F | G | H |
| UTILITIES | X | X | X | | X | X | X | |
| PUBLIC WORKS | X | X | X | | | | | |
| PLANNING | X | X | | X | X | | | |
| POLICE | X | X | | | X | | | |
| FIRE | X | X | X | X | X | | | X |
| CITY CLERK | X | X | | | | | | |
| CITY MANAGER | X | X | X | X | | | | |
| DEPT. A | X | X | | | X | | X | |
| DEPT. Z | X | X | X | | X | | | |
| PRIVATE SECTOR | X | X | | | X | X | | |
| CITIZENS . . . | X | X | | | X | | | |
| OTHER SECTORS . . . | X | X | X | | | | | |
| | UNIVERSAL USE | | BROAD USE | | | DEPARTMENTAL USE | | |

**Table 7.1** Prioritizing Web services based on internal and external needs

**Framework data and Web services are the foundation of NSDI. They serve as the geographic framework, or foundation, on which other datasets can be developed and referenced.** Framework data consists of a group of basic datasets that collectively represent the important geographic elements of a country. Framework datasets often need to cover an entire country and need to be produced with sufficient accuracy. There are no universal criteria for what themes constitute framework data (Ryerson and Aronoff 2009). **In the United States, framework datasets incorporate seven themes: geodetic control, orthoimagery, elevation, transportation, hydrography, governmental units, and cadastral information.** These themes were chosen because these types of datasets "are produced and used by most organizations. Various surveys indicate that they are required by a majority of users, form a critical foundation for the NSDI, and have widespread usefulness" (FGDC 2006). Recently, other themes have been suggested—for example, wildlife habitat or ecosystem (Dangermond and Miglarese 2009; Campbell et al. 2009), 3D urban structures (Campbell et al. 2009), and land use and soil (Craig 2009). The framework data is typically collected by many participant organizations, especially local governments, and is then aggregated by federal agencies. Making this framework data available as map and feature Web services will eliminate a lot of the costs of redundant data collecting and provide the opportunity for Web applications that will cut across government and societal needs.

In addition to the national framework services, there are many other services that offer either broad use or niche uses. These specialized services should be prioritized based on the data and expertise an organization has and the significance of market needs. Web sites that collect VGI can share it via Web services as well—for example, in GeoRSS, KML, or JSON format. Sharing VGI allows other organizations and citizens to use the information in new applications. For example, once the information on graffiti and potholes reported by citizens is shared as Web services, 2-1-1 service call centers and transportation departments can integrate this information to build workflow-related applications (figure 7.8).

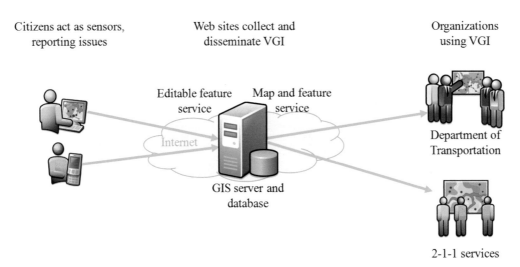

**Figure 7.8**  Issues reported by citizens can be further served out as Web services to support applications of related government agencies.

### 7.2.3 HOSTING VERSUS OUTSOURCING

Creating and hosting geospatial Web services requires GIS servers. Hosting Web services and keeping them available with high performance and scalability twenty-four hours a day requires appropriate infrastructure. Organizations that have the necessary resources can create and retain their own services—and host them for others (figure 7.9). Organizations that don't, or that don't want the expensive IT commitments, can take advantage of hosting done through a partnership or by outsourcing to companies that provide cloud computing services.

An organization with data can partner with other agencies to have them host the data as Web services. This method typically applies to organizations that don't have the necessary IT resources, but it is also part of the workflow for building framework data and services. For example, a local government can partner with upper-level government and have its data aggregated to the partner's database via real-time data replication services or periodic ETL (extract, transform, and load) processes. Upper-level government can then create and host Web services that include the data from lower-level governments (figure 7.10).

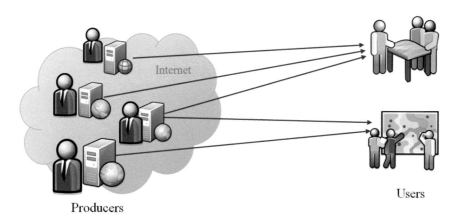

**Figure 7.9**   Organizations can serve Web services within the organization and to others if they have the necessary resources.

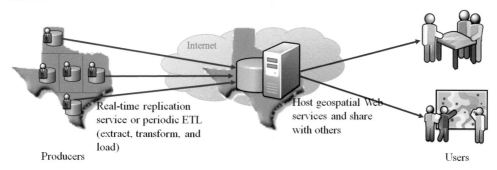

**Figure 7.10**   An organization can have its data replicated to a partner organization's database and have the partner create and host Web services.

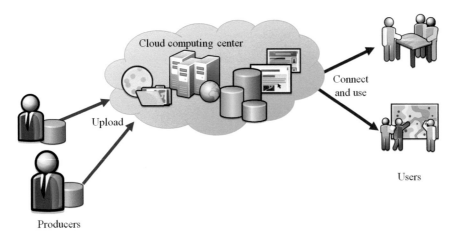

**Figure 7.11** Organizations that do not have the necessary resources to host Web services or do not want the IT commitment can upload their data and models to a cloud computing center and have their Web services hosted there.

Cloud computing is an important component of today's information infrastructure (see chapter 10). It enables on-demand access to a shared pool of configurable computing resources (e.g., networks, servers, storage, applications, and services) over the Internet. For most users, cloud computing means information-access services that come from the cloud. For others, cloud computing is a good solution for uploading data and using the cloud computing center as a hosting service (figure 7.11).

Building cloud infrastructure to host geospatial Web services is a new type of business model and is still in its early stages. For example, ESRI offers the following cloud-based solutions for hosting Web services:

- ArcGIS Online community maps program: encourages user organizations to contribute selected themes of geographic content to be published as 2D and 3D map services hosted by ESRI

- Preconfigured ArcGIS Server in the Amazon cloud: allows organizations to upload their data to the cloud, author maps and models, and publish Web services without needing local infrastructure

- ArcGIS Server for the cloud: enables more flexible hosting of Web services, allowing organizations to install their own instance of ArcGIS Server on Amazon or in their own cloud to host their own and other organizations' Web services

Once Web services are hosted in the cloud, they can be easily accessed by contributing organizations, other designated organizations, or the public. Contributing organizations can obtain continuous and reliable access to these Web services, saving money and avoiding expensive IT commitments.

## 7.2.4 ISSUES WITH TODAY'S GEOSPATIAL WEB SERVICES

For years, the GIS community has been creating, sharing, and using Web services and mashups in projects from small to enterprise scale. However, there are a number of issues associated with many of today's Web services:

- Fully functional national framework services are not yet all in place. Some layers do not cover the whole nation with consistent spatial resolution and temporal currency.

- Many Web services are difficult to consume, either because they have overly complicated interfaces or they don't comply with industry standards.

- The performance of many Web services is slow. Caching and other GIS server optimization techniques (see chapter 3) should be used.

- There is a lack of standards for cached Web services. WMS has been widely adopted, but it has also been criticized for being slow since it does not facilitate caching. The Web Map Tiled Service (WMTS) specification under review by OGC is looking at caching to facilitate fast Web maps.

- Sharing is still limited. Only a small percentage of organizations are sharing information via Web services, and most of the services being shared are map services. It would be beneficial for more organizations to participate in developing the new NSDI. While visualization is the most commonly used service, other services such as data editing, modeling, and business analysis can also be highly useful when shared with others.

## 7.3 ASSEMBLING WEB SERVICES

One of the ultimate goals of NSDI is the "democratization of data," which enables the fullest possible use of Web services to create new data products and services (Nebert 2004). In this vision of maximum use, both GIS professionals and private citizens can use lightweight mashups and user-friendly geobrowsers to assemble data and information published as Web services for diverse applications.

### 7.3.1 DISCOVERING NEEDED SERVICES

To further the goal of usability and accessibility, providers should make their Web services easy to discover. Registering Web services in portals promotes sharing across a broad spectrum. The new trend of geoportals is shifting from verbose metadata to "findability" (see section 6.4). For example, new-style metadata such as simple tags allows providers to quickly share their services. Exposing Web services metadata as HTMLs can make the services directly indexable and discoverable by general Web portals. Today's geoportals allow users to specifically search for Web services, or live data, and cloud services, and then discover, bind, preview, and use these services. For example, the General Services Administration (GSA) Web portal (Apps.gov) makes it easy for government agencies to search and purchase geospatial cloud services; the California Geospatial Clearinghouse Cal-Atlas portal (`http://www.atlas.ca.gov`) allows users to search for Web services and find the service

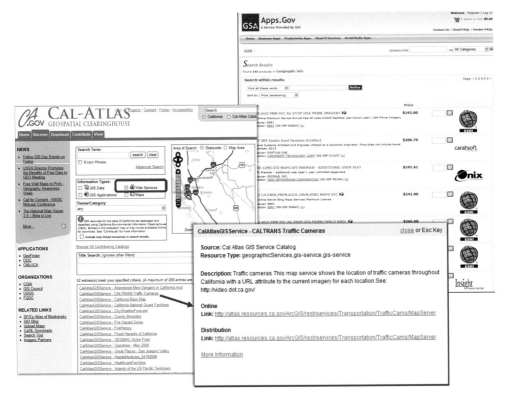

**Figure 7.12**  At upper right, the GSA Apps.gov Web portal allows government agencies to discover and use geospatial cloud services. At left, the California Geospatial Clearinghouse Cal-Atlas portal allows users to search for specific Web services (highlighted) and facilitates the preview and use of Web services.

Courtesy of the Natural Resource Agency, CERES, and U.S. General Services Administration.

endpoint URLs that lead to the service metadata and preview pages (figure 7.12); and ArcGIS.com allows users to search for various types of geospatial contents, especially Web services, to build composite maps and applications (see chapter 6).

### 7.3.2 TAKING ADVANTAGE OF CLOUD GIS SERVICES

The concept of cloud computing and cloud GIS is introduced in section 10.1.10. **Cloud GIS, or GIS equipped with cloud computing technology, has great potential.** It can provide infrastructure as a service, a GIS platform as a service, and GIS software as a service. Cloud GIS inherits the essential characteristics of cloud computing and has many advantages over traditional desktop GIS. Advantages for the user include reduced costs, since many basic services in the cloud are free or pay as you go; reduced complexity, because infrastructure is maintained by others; rapid scalability; and increased performance. A cloud can be public or private. A private cloud capitalizes on data security, corporate governance, and reliability, fitting the needs of many mission-critical or highly secured business interests.

For example, ArcGIS Online uses the cloud computing paradigm to deliver GIS capabilities in the cloud. The ArcGIS Online cloud offers a variety of contents and capabilities (figure 7.13):

- Basemap and specialized map services are available in 2D and 3D and are cached and cartographically rendered to support multiscale displays. The variety is rapidly expanding. Services currently available include imagery, streets, physical, topographic, shaded relief, navigation charts, events imagery, protected areas, historical maps, cloud cover imagery, demographic maps, and political maps. Microsoft Bing Maps are also available through ArcGIS Online.

- Analytical services, such as geocoding and gazetteer services, routing services, and a series of business analysis services (see ESRI Business Analyst Online services in chapter 8), deliver business intelligence to small-business retailers, commercial real-estate agents, and other businesses.

- ArcGIS Online supports SOA and WOA by providing REST interfaces in addition to SOAP interfaces. Web services can be mashed up using JavaScript, Flex, and Silverlight, making it easier, quicker, and cheaper to build GIS applications.

Data appliance-type products are a solution that can be used by a private cloud. For instance, the ArcGIS Data Appliance is designed for organizations that want to provide secure access to a comprehensive collection of geographic data and maps. The ArcGIS Data Appliance combines data and hardware so that it easily fits into an organization's existing IT infrastructure by connecting directly into the network. The appliance comes with terabytes of prerendered and precached data for use within the organization's own secure ArcGIS Server instance behind a firewall. The product includes most of the framework map services and basic task services of ArcGIS Online (figure 7.14). Multiple divisions, departments, and client applications can share this common

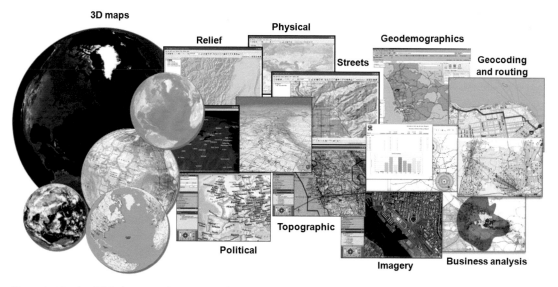

**Figure 7.13**   ArcGIS Online provides an array of 2D and 3D map services as well as analysis services in the cloud.

Map data © AND Automotive Navigation Data; courtesy of Tele Atlas North America, Inc.; GeoEye; i-Cubed; NGS; U.S. Geological Survey; NASA; and U.S. Census Bureau.

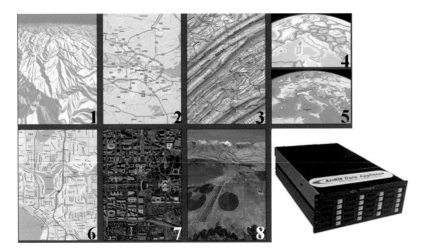

**Figure 7.14**    The ArcGIS Data Appliance includes most of the framework map services and basic task services of ArcGIS Online. Illustrated here are (1) World Shaded Relief Globe Service, (2) World Street Map Service, (3) USA Topographic Maps Service, (4) World Political Map Globe Service, (5) World Physical Map Globe Service, (6) World Street Map Service, (7) USA Prime Imagery 2D Map Service, and (8) USA Prime Imagery Globe Service.

Map data © AND Automotive Navigation Data; courtesy of Tele Atlas North America, Inc.; i-Cubed; NGS; and U.S. National Park Service.

information architecture, and a set of 2D and 3D map services can be extended by combining them with an organization's own data.

### 7.3.3 USING WEB SERVICES TO BUILD APPLICATIONS

Geospatial Web services, shared by various organizations, can be reused and remixed to build applications, which is a tenet of NSDI 2.0. The benefits have been demonstrated in many innovative and value-added mashup applications. By using the mashup concept to assemble authoritative information and robust Web services, NSDI 2.0 will support a resilient and extensive **federated GIS—that is, GIS based on a distributed collection of connected GIS nodes that share and make use of each other's geographic information and services.**

For example, the VIPER (Virginia Interoperability Picture for Emergency Response) project by the Virginia Department of Emergency Management (VDEM) is a federated GIS that illustrates the benefits promised by NSDI 2.0 as well as the challenges of a current implementation. In 2008, VDEM decided to overhaul its existing situational awareness system. The goal of the new system was to enhance information sharing, communication, and analysis; provide a new level of connectivity; fully integrate multiple systems; and remove the obstacle of wading through numerous information stores, databases, and other technologies. Preconfigured processes were put in place so that when an incident occurs, the right datasets and feeds are activated, and responders can act immediately and monitor events in real time (Atristain 2009).

VDEM identified more than a hundred Web resources from various organizations (Crumpler 2009) that would be used in the project (figure 7.15), and included many of them in the application:

- Basemaps from ArcGIS Online

- Imagery services from Virginia state government

- Information on critical facilities, infrastructure, and key resources, such as state and local shelters, pet shelters, emergency response plans, search and rescue resources, and live dispatch status from VDEM

- Real-time landslide and earthquake data from USGS

- Storm, flood, and tornado warnings from the National Weather Service

- Storm surge from the U.S. Army Corps of Engineers (USACE)

- Active fire updates from USFS

- Virginia hazards, aircraft incidents, traffic and traffic cameras, and many other services from various organizations

- Feeds from social networks such as Twitter and YouTube

Some of these resources are Web services, while others are not. The resources that are not Web services have to be dynamically retrieved from their Web sources, processed, and created as Web services using ArcGIS Server. These resources are then integrated into an ArcGIS Flex Viewer. By assembling the diverse Web resources shared by governments, the private sector, and individual

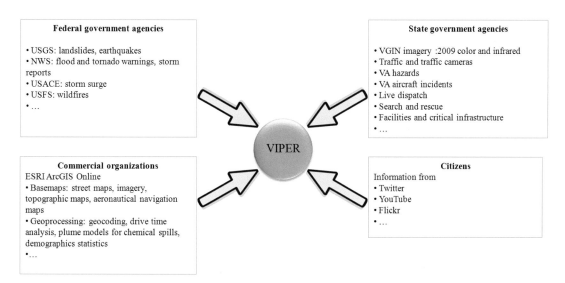

**Figure 7.15**    VIPER is built by reusing and remixing distributed Web services and resources provided by federal, state, and local governments, the private sector, and citizens to aid Virginia emergency response.

citizens, VIPER formulates a powerful common operating picture (COP) that helps VDEM and other agencies better manage and respond to emergencies:

*"VIPER provides Virginia Emergency Operations Center (VEOC) staff with the ability to visually assess statewide emergency management operations in real time. It also automatically offers instant access to essential information and maps. For example, if a particular highway experiences a severe traffic accident with multiple injuries, VIPER will provide information about nearby hospitals; in the case of a hazardous materials spill, VIPER will offer data about schools and other high-priority evacuation sites. VIPER then performs an analysis of all available information and alerts VEOC staff about potential impacts on critical infrastructure.*

*VIPER monitors environmental sensors and gathers data from VDEM's crisis management system, as well as external systems, such as Computer Aided Dispatch, the National Weather Service, and the Integrated Flood Observation and Warning System. VIPER then performs an analysis of all available information and alerts VEOC staff to potential impacts on critical infrastructure. This ability to evaluate how incidents visually relate to each other combined with point-and-click access to essential local data greatly speeds VDEM's coordination of response and recovery efforts at the state and local levels.*

*VIPER had an opportunity to prove its mettle in several high-profile natural disaster and national security events—the most notable ones being Tropical Storm Hannah, the presidential election, and the Inauguration of the 44th president of the United States, President Barack Obama. In each instance, VIPER was integral to providing geospatial intelligence and situational awareness to allied agencies supporting these events."* (Atristain 2009)

VIPER is an example of federated GIS that is built using the Web resources shared by many organizations. Some resources are not Web services and require extra processing, including periodic data downloading with specialized tools and processing with customized programs so they can be published as Web services and then integrated. The extra efforts employed by VIPER show the difficulty in reusing nonstandard Web services. Some of the other Web services used in the project, however, are RESTful Web services that can be easily integrated by simple configurations, demonstrating the benefits of the WOA principle that NSDI 2.0 endorses.

### 7.3.4 USING GEOBROWSERS TO ASSEMBLE WEB SERVICES

Geobrowsers such as ArcGIS Explorer, Gaia, Google Earth, and NASA World Wind have gained popularity in recent years. Most of them are easy, even fun, to use, and allow almost anyone, from GIS professionals to casual users, to view a map or globe, turn on additional data layers or services, and overlay services available from the Web to create new applications. Users can view high-resolution basemaps, which are Web services themselves, as well as many other services or feeds available on the Web. Examples include image services showing wildfires in national forests, GeoRSS feeds of the daily news from news agencies, geotagged photos from online albums, real-time highway traffic incidents from transportation departments, and real-time disaster

**Figure 7.16**  Geobrowsers can be used to access and assemble distributed geospatial Web services without the need for programming. In this figure, ArcGIS Explorer Desktop displays, from left, a basemap from ArcGIS Online, a real-time KML earthquake feed from USGS, and GeoRSS of a geotagged photo collection from Flickr.

KML and GeoRSS feeds from government agencies (figure 7.16). Users with advanced skills can link map services, feeds, and geoprocessing services into geobrowsers and load local data to conduct their own research.

Geobrowsers have rapidly proved their value in making geospatial data and Web services available to mass users. They foster collaboration between geospatial Web service providers and consumers; improve the alignment of geospatial products with the interests of society; and help to realize the democratization of data, the ultimate goal of NSDI.

## 7.4 CHALLENGES AND PROSPECTS

NSDI 2.0 will promote broad and pervasive partnerships between the public and private sectors and enable citizens to efficiently participate in the collecting, sharing, and use of geospatial data. Challenges must still be resolved, but a mature NSDI 2.0 will produce tremendous benefits for society.

### 7.4.1 DATA QUALITY AND SEMANTIC INTEROPERABILITY

The Web services approach to sharing information may face the classic problem of semantic heterogeneity and inconsistent accuracy in data from different sources. Semantic heterogeneity can be cognitive or in naming. Cognitive heterogeneity occurs when two organizations have a different view of the same real-world object, while naming heterogeneity means that concepts that refer to the same real-world phenomena might be named differently (Bishr 1998). As a result, the same term by different organizations might have different meanings while different terms might actually mean the same thing. Disparities can also be caused by the inconsistent spatial accuracy of different data sources. When such data is merged or overlaid, whether by data duplication or Web service integration, the discrepancies will become obvious (figure 7.17). You may see rivers that don't meet at regional boundaries, cars driving on lakes, or houses located in the middle of the street. You may also see that the boundaries of the same soil type from adjacent regions don't align uniformly because the same soil type name may mean different things in different regions, a form of semantic heterogeneity.

The accuracy of framework map services is especially important. They are often used as basemaps, on top of which VGI is marked, pinpointed, or redlined. If the framework data is not accurate, VGI and other derived data won't be accurate either. The errors can flow through an information ecosystem, being recirculated and amplified by the chain of users.

It is important for data producers to ensure data quality and semantic interoperability. And it is crucial for data consumers not to use data beyond the scope that its accuracy permits. A

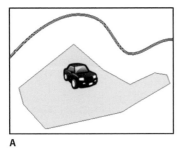
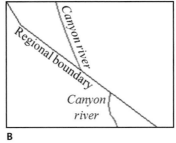
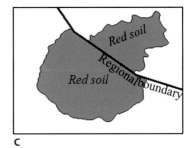

A         B         C

**Figure 7.17** Discrepancies will become obvious when data with inconsistent accuracy and semantics is brought together from different sources: (A) a car appears to drive in a lake and (B) the same river doesn't meet up correctly at regional boundaries as a result of inconsistent accuracy; and (C) the boundaries of soil types with the same name don't align uniformly at the regional boundary as a result of semantic heterogeneity.

mechanism to ensure the quality of VGI is needed for VGI to gain the same level of trust as information from authoritative sources. The importance of defining semantic interoperability standards remains a key issue for national and international standards bodies.

## 7.4.2 APPLICATION SECURITY AND RELIABILITY

Some organizations are hesitant, out of security and reliability concerns, to move all their data to Web services or to totally rely on external Web services. They choose a Software plus Services (S+S) architecture. **Software plus Services combines hosted services with the functionalities of locally running software** (Chong et al. 2009). S+S can be the appropriate choice in the following situations:

- **Offline capability:** When an application is unable to connect, or can't always be connected, to the Internet, S+S allows an application to operate locally and offline with local data. It is useful for mission-critical applications that are considered too important to risk being dependent on the Internet.

- **Security concerns:** When information is so sensitive that the consequences of it being compromised are severe, many organizations would rather keep sensitive data local and internal to their network. For example, the exposure of confidential corporate information when documents are hacked raises red flags on cloud security (Cain Miller and Stone 2009).

- **Limitations of current Web services:** While today's Web services are becoming richer, it is not realistic to expect that all functions and information that all users need be available through Web services—not now or in the future.

While advances in technology will address many of these concerns, there is still plenty of room for the S+S model. Fusion centers are a good example of S+S architecture. **The fusion center pattern is a new type of GIS that is increasingly being used to provide situational awareness. In a fusion center, multiple GIS databases and Web services are integrated into a single environment that supports applications involving real-time visualization and analysis of geospatial data** (Dangermond 2008). The S+S model is an optimal choice for situations such as when offline capability is needed, security concerns are high, or when the needed Web services are not available.

## 7.4.3 THE BUSINESS MODEL: WHO PAYS THE COST?

While users will enjoy the benefits of Web services, providers will face costs in setting them up and guaranteeing their reliability. Advances in IT will help to lower the costs, but not eliminate them. It is an expense that must be paid. Given the typical government funding model, it can be challenging to justify the expense of hosting Web services for other organizations and other projects to use.

It will take government leadership, along with adequate levels of government funding, to encourage government agencies to share their information and expertise as Web services. In the long term, the benefits of making geographic information available as services, free of charge or for reduced fees, will pay off in the form of intangible benefits. The creation of more value-added products through the synthesis of this information and services will result in increased revenues (Welle Donker 2009).

### 7.4.4 FOSTERING SMOOTHER COLLABORATION

Given the ambitious scope of NSDI, its "no one in charge" structure, and the lack of simple measures of success, it is not surprising that highly partisan views exist on what NSDI has achieved and what needs to be done next (Longley et al. 2005). However, there are few who dispute the merits of data sharing and a reduction of duplication that can result from collaboration and coordination.

The first decade of NSDI development has been primarily government-centric and oriented toward data duplication. Web 2.0 provides a suite of enhanced technologies to create the new generation of NSDI. To achieve openness and interoperability, NSDI 2.0 should be developed with SOA and WOA principles as a widely distributed system. It should build on the growing popularity of Web services, cloud computing, mashups, and geobrowsers. It should leverage VGI and engage government, businesses, academia, and citizens in collecting and sharing geospatial data. NSDI 2.0 must be designed to adequately accommodate the market-oriented realities of the geospatial industry and ensure the alignment of business objectives with community interests. NSDI 2.0 can deliver the means for flexible, interoperable, real-time collaboration on the use of geospatial information to an extent that NSDI 1.0 has not.

Despite the challenges, the benefits of an emerging NSDI 2.0 are clearly demonstrated in the positive results of many successful projects. NSDI 2.0 can provide the foundation for the development of e-business (see chapter 8) and e-government (see chapter 9), and promises an exciting array of new opportunities for the use of GIS within the society as a whole.

---

### Study questions

1. What is NSDI? Why is sharing information important to NSDI?

2. What are the main characteristics of NSDI 1.0? What about NSDI 2.0?

3. What forms have been used to share geospatial information? Compare the disadvantages and advantages of these methods.

4. List the Web 2.0 technologies that will benefit NSDI 2.0 and describe the technical philosophy of NSDI 2.0.

5. What are SOA and WOA?

6. Give some examples of how NSDI 2.0 can involve citizens in data production.

7. Explain the alternatives an organization can use to publish and host its data, maps, and models as Web services.

8. What issues do today's geospatial Web services pose for the realization of NSDI 2.0?

9. Describe a GIS cloud, including its contents, functions, and programming interfaces.

10. What problems might you run into when combining geospatial Web services from different sources?

11. What is the Software plus Services architecture? When is it needed?

12. What geospatial contents and functions do you or your organization have that are needed by other organizations? How can they be shared?

13. What geospatial Web services offered by other organizations do you wish you had to enhance your job or your personal life?

14. Try to integrate multiple geospatial Web services or feeds using a geobrowser.

## References

Atristain, Bobbie. 2009. VIPER: Virginia deploys Web-based emergency management system. *ArcNews* (Fall). `http://www.esri.com/news/arcnews/fall09articles/viper.html` (accessed November 18, 2009).

Binns, Andrew. 2007. Reengineering SDI development to support a spatially enabled society. In *Towards a spatially enabled society,* ed. Abbas Rajabifard. Melbourne, Australia: University of Melbourne.

Bishr, Yaser. 1998. Overcoming the semantic and other barriers to GIS interoperability. *International Journal of Geographical Information Science* 12 (4): 299–314.

Box, Paul, and Abbas Rajabifard. 2009. SDI governance bridging the gap between people and geospatial resources. GSDI 11 World Conference, Rotterdam, The Netherlands, June 15–19. `http://www.gsdi.org/gsdiconf/gsdi11/papers/pdf/151.pdf` (accessed November 16, 2009).

Cain Miller, Claire, and Brad Stone. 2009. Hacker exposes private Twitter documents. *New York Times*, July 15. `http://bits.blogs.nytimes.com/2009/07/15/hacker-exposes-private-twitter-documents/?hp` (accessed November 30, 2009).

Campbell, Lisa, John Curlander, Steven R. Hagan, Michael T. Jones, and Jack Pellicci. 2009. A proposal for reinvigorating the American economy through investment in the U.S. National Spatial Data Infrastructure (NSDI). `http://www.cast.uark.edu/research/projects/nsdi/nsdiplan.pdf` (accessed November 28, 2009).

Craig, Will. 2009. Governance of the NSDI. *ArcNews* 31 (3). `http://www.esri.com/news/arcnews/fall09articles/governance-of-nsdi.html` (accessed November 15, 2009).

Chong, Fred, Alejandro Miguel, Jason Hogg, Ulrich Homann, Brant Zwiefel, Danny Garber, Joshy Joseph, Scott Zimmerman, and Stephen Kaufman. 2009. Design considerations for S+S and cloud computing. *Architecture Journal*, September. `http://msdn.microsoft.com/en-us/architecture/aa699439.aspx` (accessed November 28, 2009).

Crumpler, Brian. 2009. Establishing a geospatial information management framework for emergency management in the Commonwealth of Virginia. ESRI Southeast User Group Conference, Jacksonville, Fla., April 27?29.

Dangermond, Jack. 2008. The work of GIS professionals and evolving GIS technology: GIS—geography in action. *ArcNews* (Winter). `http://www.esri.com/news/ArcNews/winter0809articles/gis-geography-in-action.html` (accessed November 27, 2009).

Dangermond, Jack, and Anne Hale Miglarese. 2009. An investment in geospatial information infrastructure: Building a national GIS. `http://www.nysgis.org/Docs/National_GIS_proposal_1-09.pdf` (accessed November 28, 2009).

Dangermond, Jack, and Jeremy Bartley. 2009. Geoenabling Gov 2.0. Gov 2.0 Summit: The Platform for Change, Washington, D.C., September 9?10.

FGDC. 2006. Framework frequently asked questions (FAQs). `http://www.fgdc.gov/framework/frameworkfaq#q4` (accessed December 1, 2009).

Gall, Nick. 2008. *WOA: Putting the Web back in Web services*. Stamford, Conn.: Gartner. `http://blogs.gartner.com/nick_gall/2008/11/19/woa-putting-the-web-back-in-web-services` (accessed November 19, 2009).

Haas, Hugo. 2003. *Designing the architecture for Web services*. WWW2003, Budapest, Hungary. `http://www.w3.org/2003/Talks/0521-hh-wsa/slide5-0.html`.

Harrison, Jeff, John Moeller, Julia Harrell, Jason Mann, Edric Keighan, Steven Johnson, Julie Stamper, David G. Smith, and Bill Dollins. 2009. NSDI 2.0: Powering our national economy, renewing our infrastructure, protecting our environment. http://www.nsdi2.net/NSDI2ConceptForAmericanRecoveryAndReinvestment_V1_3.pdf (accessed November 25, 2009).

Hinchcliffe, Dion. 2008. What is WOA? It's the future of service-oriented architecture (SOA). http://hinchcliffe.org/archive/2008/02/27/16617.aspx (accessed November 19, 2009).

Kash, Wyatt. 2009. GCN interview of Jack Dangermond: The next step for agency GIS; Shared services. *Government Computer News*, November 5. http://gcn.com/Articles/2009/11/09/Interview-ESRI-Jack-Dangermond.aspx?Page=1 (accessed November 15, 2009).

Kralidis, Athanasios Tom. 2007. Geospatial Web services: The evolution of geospatial data infrastructure. In *The geospatial Web: How geobrowsers, social software, and the Web 2.0 are shaping the network society*, ed. A. Scharl and K. Tochtermann. London: Springer.

Longley, Paul A., Michael F. Goodchild, David J. Maguire, and David W. Rhind. 2005. *Geographic information systems and science*. 2nd ed. San Francisco: John Wiley & Sons.

Morehouse, Scott, Dirk Gorter, and Brenda Wolfe. 2009. Web GIS taking advantage of the cloud. ESRI International User Conference, San Diego, Calif., July 13–7.

National Research Council. 1993. *Toward a coordinated spatial data infrastructure of the nation*. Washington, D.C.: National Academies Press.

Nebert, Douglas D., ed. 2004. *Developing spatial data infrastructures: The SDI cookbook*. http://www.gsdi.org/docs2004/Cookbook/cookbookV2.0.pdf (accessed September 16, 2009).

Ryerson, Robert A., and Stan Aronoff. 2009. Geospatial framework data: Definitions, benefits, and recommended data sources for equatorial regions. http://www.geosar.com/downloads/WhitePaper_GeospatialFramework_10-2009.pdf (accessed December 3, 2009).

Schulte, Roy W., and Yefim V. Natis. 1996. Service-oriented architectures, SSA Research Note SPA-401-068. Stamford, Conn.: Gartner.

The White House. 1994. Executive Order 12906: Coordinating geographic data acquisition and access: The National Spatial Data Infrastructure. http://govinfo.library.unt.edu/npr/library/direct/orders/20fa.html (accessed November 11, 2009).

Thies, Gunnar, and Gottfried Vossen. 2008. Web-oriented architectures: On the impact of Web 2.0 on service-oriented architectures. IEEE Asia-Pacific Services Computing Conference, 1075–82.

Welle Donker, Frederika. 2009. Public sector GeoWeb services: Which business models will pay for a free lunch? In *SDI convergence, research, emerging trends, and critical assessment*, ed. Bastiaan van Loenen, Jaap Besemer, and Jaap Zevenbergen. Rotterdam, The Netherlands: Optima Graphic Communication.

Wu, Lun, and Tong QingXi. 2008. Framework and development of digital China. Science in China Series E, *Technological Sciences* 51: 1–5.

Zhao, Peisheng, Genong Yu, and Liping Di. 2007. Geospatial Web services. In *Emerging spatial information systems and applications*, ed. B. N. Hilton, 1–35. Hershey, Pa.: IDEA Group.

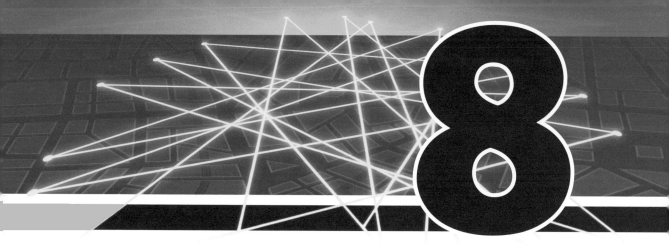

# WEB GIS APPLICATIONS IN E-BUSINESS

Geography applications are longtime benefactors of business operations. Web GIS is instrumental in generating tremendous revenues from existing online businesses and has spurred the development of innovative new Web-based businesses. Because Web maps are an extremely effective way to deliver business listings, online advertising companies have adopted them as a replacement for traditional directory, or Yellow Pages–style, listings. The use of Web maps to find stores, restaurants, hotels, addresses, and optimal routes is nearly universal. In addition, Web GIS itself can be a commodity in the form of Software as a Service (SaaS) or cloud-based GIS. Organizations that need various types of GIS functionality can purchase the specific GIS data, maps, and analytic capabilities they need. Because GIS is now widely available and more accessible to a broader audience, more businesses are adopting it to manage customer relations, select profitable business locations, and support business decisions.

This chapter discusses how Web GIS relates to electronic business (e-business). The first section provides information on the nature and development of e-business alongside a review of applications of GIS in business. The second section introduces Web GIS applications that aid online advertising, facilitate business operations, and support business analysis. It also illustrates how Web GIS can be delivered as a commodity via the SaaS business model. The last section describes the challenges and prospects for Web GIS applications in e-business.

## 8.1 INTRODUCTION TO E-BUSINESS AND GEOBUSINESS

Since the 1990s, the Web has transformed how most businesses operate. Although the volume of retail sales on the Web has climbed steadily, the impact and scope of e-business far exceeds the growth in online sales. In addition to changing how companies sell products and services, the introduction of **the Web has changed how businesses advertise, manage employees, support customers, interact with suppliers, and make decisions.**

### 8.1.1 THE EMERGENCE AND DEVELOPMENT OF E-BUSINESS

The broadest and most basic definition of **e-business is the use of information and communication technologies (ICT) and the Web for all types of business activities, ranging from advertising to supply chain management to merchandising.**

Discussions of e-business often ignore the big picture and focus narrowly on e-commerce, the online sale of merchandise, which is a highly visible subset of e-business. Although online sales might come most readily to mind, that definition is far too restrictive. **E-business encompasses the use of Internet protocols and related technologies to benefit business-to-business (B2B), business-to-employee (B2E), business-to-government (B2G), and business-to-customer (B2C) interactions.** The introduction of e-business has profoundly transformed the business landscape, changing what companies do and how they operate.

The term e-business gained common currency via an IBM advertising campaign in the 1990s that emphasized a shift in the company's technology toward e-business and promised "the transformation of key business processes through the use of Internet technologies" (Li 2007).

The U.S. government also quickly recognized the potential of the Internet to revitalize the economy. In 1995, President Clinton described the Internet as an "engine for future economic growth." He proposed an e-commerce policy, the Framework for Global Commerce, to encourage e-business ventures, which became the Internet Tax Freedom Act (ITFA). In effect from 1998 to 2001, the act prohibited the imposition of new e-commerce taxation.

Events have largely borne out the assertion, made in 1999 by then-Intel president Andy Grove, that within five years' time, all companies would be Internet companies, or they wouldn't be companies at all (Symonds 1999). This meant not just B2C online consumer sales, but B2B sales between companies. Many businesses enthusiastically adopted e-business because the Internet promised such a pronounced competitive advantage. In addition to the cost-cutting benefits associated with carrying out transactions, e-business offered scalability, rapid implementation and modification of infrastructure, and the ability to implement functionality in stages.

In the intervening years, e-business has lived up to these early prognostications. According to a 2009 study, "Economic Value of the Advertising-Supported Internet Ecosystem," commissioned by the Interactive Advertising Bureau and carried out by Harvard Business School professors John Deighton and John Quelch and Hamilton Consultants of Cambridge, Massachusetts, the direct economic value of services provided by the Internet to the rest of the U.S. economy is estimated at $175 billion (2009). The study found that the Internet directly employs 1.2 million people, who build or maintain Internet infrastructure, facilitate its use, or conduct advertising and commerce on the Internet.

The report notes:

*"In two decades, the Internet has become central to social and economic life and is, today, a mature and integral element of the U.S. national economy. It is not only vital infrastructure, it is a spur to entrepreneurship and social change."*

**The business models for e-business can be broadly classified as businesses in which some, many, or nearly all aspects of operations have migrated to the Web and native businesses that are unique to the Web.** E-businesses in the first category use the Internet for activities such as catalog sales, advertising, offering free trials, direct marketing, incentive marketing, and supplying services such as banking or stock trading. In contrast, native e-businesses typically were started as free products or services that evolved and became businesses. These businesses offer digital products and services that are delivered over the Web (Bambury 2006). **The Web has spawned at least three new native business models: aggregators, online auctioneers, and exchanges.** Aggregators provide a single source of current product and price information for markets that can range from agricultural commodities (e.g., National Spot Exchange Limited) to rare wines (e.g., Vinfolio Marketplace). Online auctioneers such as eBay, which allow sellers to peddle their merchandise, are entrenched in the public consciousness. Online exchanges, such as WorldEnergy which allow for the sale of commodities, add fluidity to markets—in this case, the energy, green power, and environmental commodities markets.

The native-business category is the larger of the two and continues to grow rapidly as businesses realize that having a Web presence is not just an option in a competitive marketplace. For these "transplanted" businesses, e-business is not a new type of business, but a new way of doing what they have always been doing. The Web augments and enhances traditional business operations and can increase the reach of a business tremendously. Unlike traditional advertising done in telephone books or on billboards, where it costs more to reach more people, there is no additional cost to reach new customers via online advertising, aside from the threshold cost. E-business works to supplement sales by allowing customers to purchase items online as well as by mail and "in store." And it supplements advertising by allowing customers to use the Internet to find out store hours and locations and to evaluate products and services. Companies are no longer limited to offering only items that are in stock or at a particular store location. With e-business, many more items can be sold and sent directly from supplier to customer.

**Speed and convenience are the two defining characteristics of Web-based business interactions, and this is particularly true of e-commerce.** In most cases, the online shopping experience is essentially catalog shopping with 24/7 availability and enhancements such as the ability to quickly locate an item, determine whether it is in stock, and immediately order and pay for it. By adding an e-mail promotion containing a link on the Web site to a sales item, an e-business can notify potential customers more quickly than with a mailer, and recipients can gain the satisfaction of buying the item immediately. Indeed, online shopping takes impulse buying to a new level, with shopping available anywhere on any device with an Internet connection.

E-business sites typically employ additional indirect advertising strategies that differ from the standard brick-and-mortar approach to selling goods and services. Web sites influence customer decisions by offering general information on related topics in addition to information on a specific product. An engaging and intuitive user interface design is essential for attracting and keeping customers on the site. Some sites promote online communities and offer games and contests to encourage customers to return to the site regularly. Ensuring that customers visit and revisit the site is important for another reason—advertisements by other companies on the site can generate additional revenue streams for e-business.

The flexibility of the Web introduces a degree of nimbleness to marketing efforts that goes beyond traditional print advertising. Marketing messages and offers can be quickly modified in online advertising. Interactions online can replace interactions with a company's employees, reducing the cost of overhead, and a sophisticated Web presence can create a positive impression of a company at lower cost.

**E-business has also restructured the workflow and B2E relations for many companies.** Intranets can make current information instantly available and are used as the conduit for routine interactions between human resources departments and staff. Web technologies enable employees to work offsite with access to the materials and information they need. The same Web technologies that allow employees to work remotely also enable companies to outsource staffing when needed.

B2B and B2G transactions such as sourcing and tendering, supply chain management, and review and approval processes are handled easily online. Web conferencing and other Web-based collaboration platforms let managers and employees interact with customers and business partners without having to spend time or money traveling.

## 8.1.2 THE VALUE OF GIS TO BUSINESS

Businesses need answers to questions that start with "Where?" Where are the best customers located? Where should new stores be located? Where should sales territories be located? GIS technology answers these questions more effectively than any other information technology by integrating a wide variety of geographic and business-related data, interacting with other information technology software, and visualizing these answers in an easily understood format—a map. Applications of GIS in business, or geobusiness (Pick 2008), are more than electronic substitutes for pushpin maps. They are important analysis and decision support tools that businesses rely on for day-to-day operations. GIS helps businesses improve product sales and distribution by focusing on marketing efforts, routing service personnel, and monitoring and analyzing productivity.

Businesses can use GIS to add a spatial dimension to data to perceive patterns and relationships that aren't as apparent from static tables and charts. Data on demographics, business locations, and other information specific to industries such as finance or real estate can be obtained from commercial sources at a reasonable cost and integrated with customer data.

Industries such as utilities and telecommunications that use GIS to manage facilities and materials are finding that the use of GIS can be expanded into areas of business operations such as marketing and customer service. Insurance, finance, and health-care companies are also making use of GIS. Insurance companies use GIS to manage their exposure to risk and estimate costs to accurately determine the price of premiums. Banks and financial institutions that began using GIS to fulfill Community Reinvestment Act (CRA) requirements for identifying redevelopment areas have found that they can use GIS for target marketing, goal planning, facilities management, and consolidation of assets following mergers and acquisitions. Hospitals are using GIS applications to manage finance and human resources, aid in marketing, improve operations and customer service, and help coordinate research and planning efforts.

The connection between retail sales and GIS is intuitive. Retail businesses use GIS in market analysis to see where their best customers are located, where potential customers with similar characteristics are located, and where underserved markets are located. They use GIS to see where competitive threats exist, and where suppliers and employees are located relative to existing and planned stores. Deciding to open a new store is a big commitment for a company. Before devoting time and money to new locations, businesses run reports that identify key variables that contribute to the success of existing stores and locate potential sites with similar characteristics. In addition to considering the locations of competitors in relation to proposed sites, a business can model the effect of a new site on existing locations so that adjustments can be made to prevent a new store from cannibalizing the market area of existing stores. Similarly, GIS can balance and determine the optimal configuration of sales territories, that is, the geographical area for a sales unit such as a store or a distributor.

Sophisticated analysis incorporating GIS can intelligently direct marketing efforts by combining customer data with commercially available socioeconomic data to reveal new clients and potential markets for new products. For businesses such as telecommunications, expanding markets entails a substantial commitment of resources because it requires the creation of additional physical infrastructure to serve customers.

The value of the data integration and analysis tools provided by GIS has become vital to the reinsurance industry. Reinsurance companies sell insurance to insurance companies to protect against the losses from large natural and manmade disasters. By using GIS to analyze the risks of current and potential holdings in their portfolios, reinsurance companies ensure that they do not have excessive exposure in any one area and can accurately price reinsurance coverage. In the late 1990s and early 2000s, analysts at Partner Reinsurance Co. Ltd. used customized desktop GIS software to assign risk by county or ZIP Code and break that risk down by business type (e.g., automobile, household, commercial). For example, to determine the risk from hurricanes, analysts overlaid historical hurricane paths on the area that the client wished to cover with a reinsurance policy. The expected losses were calculated and used in the creation of a loss curve that could be incorporated into the model to determine how much to charge for the reinsurance policy.

**The growing incorporation of GIS in the decision-making process for business has led to the integration of GIS with business intelligence (BI) software.** Much like GIS, BI software provides decision support by marshaling detailed and disparate information into a coherent strategy. Originally developed to give Fortune 500 companies a competitive edge, BI has become a multibillion-dollar industry in its own right. It has evolved from early applications that furnished simple reporting and querying capabilities to data warehousing; data, Web, and text mining; online analytical processing (OLAP); and data visualization. The financial services, banking, insurance, manufacturing, telecommunications, and health-care industries adopted BI early on, but other sectors are not far behind in realizing the benefits of this technology.

BI tools are applied in risk analysis, customer analysis, site selection, territory management, market analysis, and customer relationship management (CRM). These tools excel at extracting the "who", "what," and "when" aspects of data, but are lost when it comes to the "where" element—despite the fact that businesses typically collect an array of addresses relating to each customer. The lack of "where" in BI software has been more of a conceptual flaw—the failure to see the value of locational information beyond its place on a mailing label—than the lack of spatial dimension in data. Location provides a framework for understanding relationships within datasets and between different types of data. GIS professionals have long been aware of this fact, but the business community, and BI software users specifically, have only recently appreciated the value of integrating these technologies. Combining BI with GIS adds new analytical capabilities. The "where" component makes other kinds of data accessible. Data on demographics and consumer spending can be joined with in-house data based on location to provide a more profound look at customers. BI and GIS are complementary, rather than competing technologies. Used together, they can elicit additional information and make better decisions possible.

The old saying "seeing is believing" should perhaps be replaced by the more accurate "seeing is understanding." When they have deep understanding, they are said to have "insight." People are visually oriented. BI software users needed a more robust method for visualizing and comprehending data. Charts and graphs, the standard visualization aids, are often not effective in conveying information at the level of detail commonly generated by BI tools. Attempts to synthesize this complexity into a chart can be ineffective and may simply overwhelm rather than enlighten. The ability to interrogate charts and graphs to uncover other aspects of data is also limited without a geographic context.

Many of the questions BI seeks to answer, such as exploring the demand for a product or service in an area or deciding where to locate a store, are intrinsically geographic activities. GIS software can incorporate many layers of data into an analysis and use powerful tools, such as grid surfaces, that translate complex data into useful and understandable representations that incorporate a variety of factors. Using GIS to visualize the results of BI analysis is vastly more powerful than using a graphic representation or CAD drawing since they lack a real-world context. GIS relates data to a specific location on Earth. Only GIS software can use the results of data exploration and analysis and apply those results to optimize activities such as delivering products or routing service personnel.

Large retail stores have realized this maxim and gained a competitive edge in a very competitive market. Combining GIS and BI allows these companies to manage repair technicians, delivery personnel, and others who must visit a vast number of homes each year. Intelligently handling this workforce entails more than simply routing service vans. BI combined with GIS has helped retailers forecast demand and allocate resources, improve technician productivity, reduce overtime, and lower vehicle expenses per stop.

The use of a map to easily condense a wealth of information into a visual image is obvious to anyone who has ever tried to give detailed directions. Typically, about a paragraph into the narrative needed to guide someone to a store or house, the person giving directions will become exasperated, grab a scrap of paper, and start sketching a map. Maps provide a practical, economical way to make detailed and interrelated information quickly comprehensible. Adding maps to BI brings a new competitive advantage that allows decisions to be based on multiple data sources, and then communicates the resulting information in a way that is readily understood.

For companies that have several BI packages in place, GIS can bring functionality to a range of BI products. Putting the "where" into BI has the potential to cut costs associated with administration, procurement, and operations while limiting risk. The benefits of using GIS technology to derive greater efficiency from automating or improving workflows, which are familiar to longtime GIS users, are now available to BI users who integrate GIS into their operations. The enhanced communication made possible through GIS visualization and analysis capabilities gives a business a common operating perspective that greatly aids decision making.

Before the advent of Web GIS, all GIS business applications were developed for use by specialists working on desktop computers. These applications were successful in letting businesses integrate data sources, answer questions, and optimize operations in ways that were not possible before. Still, the use of GIS was limited, because these applications were typically used by specialists who had domain knowledge and GIS expertise and who worked in one department. All GIS requests were funneled through these specialists, which imposed a burden on GIS staff who could handle only a finite number of requests. Investments in GIS staffing, training, hardware, data, and software had to be continually justified to business management. In addition, GIS was underused, because staff, particularly executives, might be only dimly aware of the potential applications of GIS technology and of the benefits it could furnish.

**Web GIS has profoundly changed the use of GIS for business by making GIS technology far more accessible. The use of intuitive, focused Web-based applications has eliminated the need to locally install special hardware, data, and software to facilitate the use of Web GIS by business.** Simple Web GIS applications do not require excessive training and can be used by staff throughout an organization, from line workers to executives. ESRI Business Analyst Online (BAO), an SaaS application, allows businesses to use online GIS analysis tools with an extensive library of demographic, consumer spending, and business data to generate specific reports and maps to answer business needs. Businesses can purchase a single report or map or subscribe annually to have unlimited access to these services and data. Web GIS, made available through mobile devices, lets staff act more quickly and effectively with data available in the field. Mobile GIS applications have been developed that optimize field inspections and surveys, incident reporting, and tracking, promoting collaboration. Mobile GIS applications that let consumers use location-based services (LBS) to integrate their current locaton with the location of nearby businesses or other attractions helps businesses generate revenue (see chapter 5).

## 8.2 TYPES OF APPLICATIONS

Businesses have found many ways to use Web GIS. This section discusses several main types of business applications that use Web GIS: online advertising, business operations, business analysis, and applications of LBS. The integration of GIS and BI technologies opens new opportunities for leveraging investments in existing data as well as in existing systems such as enterprise resource planning (ERP) and CRM. This section also shows how Web GIS itself can be a commodity and provided as services that can be purchased by subscription.

## 8.2.1 SPATIALLY TARGETED ONLINE ADVERTISING

The Web has made location more important than ever for online advertisers and marketers. For products that can be delivered by mail or be downloaded online, the location of the seller is not important. But a large percentage of online searches involve personal visits, such as driving to a restaurant for dinner or having a plumber come over to fix a water pipe. For such situations, the geographic proximity of a person to a business, whether to a plumber or a restaurant, is one of the most important dimensions of an online search. Whereas keyword searches that lack a geographic reference return hundreds of thousands of results, they are typically not helpful unless they are ranked by location. The person needs to find a plumber that serves his or her area or a restaurant that is located within a few miles.

Consumer Web mapping applications and search portals typically depend on advertising for revenue. To support spatially based online advertising, advertising companies often use the geo-targeting method—that is, determining the physical location of a Web visitor and delivering tailored content based on the visitor's location (see section 10.1.6). In order to support geotargeting, the advertising companies need to have at least two types of spatial data: the locations of businesses and the area the user wishes to search.

- **The locations of businesses:** A business location or **point of interest (POI) is a location that a user may find interesting or useful.** POIs can be locations of schools, auto repair shops, entertainment venues, parking lots, hotels, gas stations, or homes for sale (figure 8.1). Information for each POI usually includes its name, type of business, description, latitude and longitude, street address, and telephone number. Other information, such as customer rating, photos, and Web site links, is also desired for certain types of POIs. Advertising companies can collect information about POIs in several ways: by sending crews equipped with GPS receivers into the field, by geocoding businesses based on address listings in telephone directories, and by crawling and parsing Web pages for businesses. Companies placing the advertisements may also supply location information.

- **The area the user wishes to search:** This information can be obtained by an online advertising company either directly or indirectly. The user can directly supply this information via address, place-name, or ZIP Code. The search engine then parses this information to derive the area of interest.

Depending on the area a user wishes to search, the user's current location may or may not be relevant to a business. For a user in one city who is looking for a place to stay when visiting another city, their current location is inconsequential. A user's current location becomes more relevant when the user does not specify a search area. When a user doesn't specify location information, an advertising Web site typically assumes the user is more interested in businesses close to the user's current location—for instance, the user won't go too far for a dinner.

The Web site can elicit a user's current location in a number of ways:

- By the IP address of the user's computer.

- By a GPS unit in a cell phone or other device.

- By the location information the user provided at registration. Social networking sites and Web e-mail sites have users register and can use this method to deliver ads pertaining to the user.

Online advertisers such as Zillow use the location users specify to find prices of real estate based in that area. This site is all about location, location, location—in this case, the location of the property to be valued. Zillow geocodes the property of interest and uses public and commercially available data sources on recent sales and listings in the area. It then uses a mathematical model to generate a "zestimate"—that is, and approximate valuation of the property. The site offers comparable listings and sales of homes along with demographic information for local neighborhoods. The site is funded by advertising that includes mortgage companies that lend in the area.

Researchers have explored ways to determine location through information implicit in the search query. A paper titled "Discovering Users' Specific Geo Intention in Web Search" (Yi, Raghavan, and Leggetter 2009) explores the use of an analysis program that attempts to determine if individual queries are related to geography. The authors' research revealed that 13 percent of searches involve some level of geography. Of those searches, half did not include a location in the query. Of queries with location constraints, 84 percent were seeking locations at the city level. The researchers used language modeling to discover the geographic constraints implicit in a query. For example, someone ordering a pizza or looking for a coffee shop will most likely want a store within the same city and probably within the same neighborhood.

Some advertisers use adware and spyware to gather information about users. These programs are often installed unintentionally by Internet users. Such programs operate without an Internet

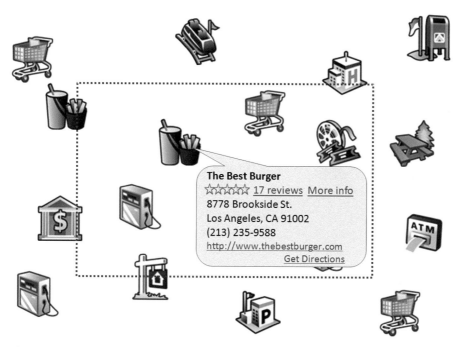

**Figure 8.1** POIs can include banks, gas stations, stores, ATMs, and entertainment attractions. Targeted online advertising is accomplished by making a spatial query of one or more types of POI that fall within the area defined by the search (indicated by the red dashed line). Users can hover over POIs to find relevant information.

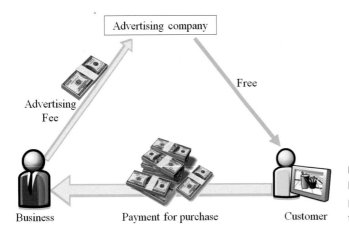

**Figure 8.2** In the online advertising business model, advertisers pay a fee, but potential customers can obtain information for free.

user's knowledge or consent and employ the user's Internet connection as a back channel for delivering personal information. These programs often collect a user's online surfing habits (e.g., Web sites visited and keywords searched) and other private information so advertising companies can deliver more relevant ads to the user.

The targeted online advertising business model, used by companies like Google and Microsoft, typically makes POI information available at no charge to the user (and potential customer). The assumption is that this information will lead to purchases from other companies identified by the application (figure 8.2). Companies pay the portal or online advertising firm to have their information included in POI listings or to have it displayed prominently to make it more noticeable.

Based on that business model, a successful online advertising site has the following characteristics:

- Is free to Web users
- Has a large collection of POIs to attract a diversity of users
- Has a large user base that will generate traffic to the site
- Focuses on the user experience and is fun and easy to use

## 8.2.2 BUSINESS OPERATIONS

Daily business operations involve a wide range of activities, including customer relationship management (CRM) and enterprise resource planning (ERP).

CRM is a broadly recognized strategy for managing and nurturing a company's interactions with customers and sales prospects. It involves using technology to organize, automate, and synchronize business processes—principally sales-related activities, but it also is used for marketing, customer service, and technical support. CRM is more than software. In fact, early incarnations of CRM systems were not software based, but salesperson based. Sales staff knew their customers and their habits and suggested products based on this knowledge. CRM software systems were created as a way to capture that knowledge; incorporate metrics to measure performance; and

**Figure 8.3**   In this mockup application, Web GIS helps a water company to quickly fix a break in a water main by identifying the valves to close and dispatching field crews to the job. The application also lists the customers that will be affected so that they have the right expectations about when the problem will be fixed.

Courtesy of Tele Atlas North America, Inc., and used with permission of Microsoft.

to carry out analyses that would anticipate, and perhaps even trigger, customer behavior. More mature CRM systems use technology to gain and retain customers by organizing and automating business processes in relation to customer interactions across the business, from marketing to customer service.

Web GIS can support business operations, and CRM activities, in many ways. For example, most retail Web sites let you find stores nearby by typing a city name or ZIP Code. The site may also display a map showing all store locations in the area and supply tools to route a customer from home to store.

Web GIS can make businesses more responsive to customer needs. For example, Web GIS can help a water company react more effectively to a broken water main, all at the click of a button. Call center operators, who are not GIS experts, can use a Web application to map the location of the break, identify the valves that need to be closed, and determine which customers will be affected (figure 8.3). An operator can send information to field crews via mobile phones. These field crews can use mobile maps and routing information to find the source of the problem. And operators can keep affected customers updated on when the problem is expected to be fixed.

FedEx makes extensive use of Web GIS to satisfy its high-end customers. Custom Critical Solutions, a special services division of FedEx, uses GIS to meet its customers' exacting requirements for timely arrival, temperature control, secure transportation, and other special shipping needs, such as handling hazardous materials or taking extra care in handling art and high-value vehicles. The needs of this FedEx division far exceed the typical dispatch and route planning technologies supplied by most navigation software. Vehicles are tracked and the condition of shipments monitored in real time by on-board sensors. Routes can be reconfigured on the fly to react to changes in road conditions or in the condition of the shipment, which is particularly important for perishable items. If necessary, vehicles can be sent for repair en route. Geofencing applications automatically alert customers when shipments are within a stipulated proximity of their destinations. This highly customized and responsive system helps FedEx to use the company's fleet optimally and keep its high-value customers satisfied.

All these operations can be done within focused Web GIS applications that are easy to use and don't require users to be experts in GIS. Web GIS can help companies get to the root of problems quickly and find solutions that build and maintain good customer relationships.

Enterprise resource planning (ERP) is an enterprise-wide system to improve and streamline internal business processes. ERP combines various software information systems to integrate activities across company departments (e.g., accounting, product distribution, inventory control). Like CRM, another enterprise-driven system, ERP is best if properly integrated with other information systems such as GIS.

Many major industries, including insurance, health, and other businesses, integrate GIS with ERP systems to manage major business functions on an enterprise level. Although originally developed for use in manufacturing, ERP software is used by all types of businesses for a variety of business operations, including human resources management, supply chain management, warehouse management, and digital dashboards that help executives track corporate functions. SAP, a pioneer in ERP software, uses ESRI ArcGIS software to integrate GIS with ERP processes to streamline operations (figure 8.4). The exact location of a problem can be determined with the use of GIS and then be used to generate an SAP maintenance request or create a list of customers who will be affected. GIS, available through the Web viewer, is the glue that holds the other information systems together.

Pidpa, a Belgian water utility and one of the largest providers of drinking water in the Flanders region, uses many information systems in addition to GIS, including ERP, supervisory control and data acquisition (SCADA), and Laboratory Information Management System (LIMS), to operate

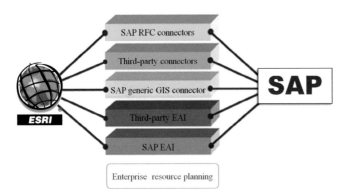

**Figure 8.4** SAP, a main provider of ERP software, and ESRI and its business partners have developed extensive interfaces that integrate GIS and ERP software systems.

and monitor twenty-six wells located in a service area of 2,581 square kilometers. Water distribution would seem like a simple business, but it requires constant vigilance and strict adherence to quality standards. Water in the network must be checked on a daily basis by the company's labs. If a problem arises, it must be contained, and alternate water must be distributed to affected customers.

Capturing, maintaining, and using the information related to management of Pidpa's water distribution, quality control, customer service, and sustainable water resources is a complex task. The company has moved from in-house development of proprietary information systems to an ERP model that uses standard software systems with limited programming requirements.

Implementing ERP throughout the company was instrumental to Pidpa's IT strategy. An intranet GIS viewer was developed to integrate GIS with SAP, SCADA, and other IT systems used within the organization. Though originally devised for GIS editing tasks, the viewer now gives access to geographic information to everyone in the company. The GIS viewer can be tailored to provide different types of information depending on the task. For example, the viewer can provide an overview map showing water production centers, water towers, and pumping stations to workers involved in maintaining the distribution network.

Many types of utilities combine ERP and GIS technologies, including oil and gas companies, real estate companies, forestry and forest products companies, airports, and service providers who need routing and logistics capabilities. The extent of integration of GIS with ERP has also increased from initial simple mapping and read-only access to data maintained in SAP to embedded GIS functionality that allows two-way interaction.

Businesses use ERP systems to improve decision making. Managers need enough information at hand to develop and evaluate solutions. The computerization of business processes allows for better data capture and analysis to aid decision making. Each type of manager, from financial to operations to marketing to human resources to administrative, has specific information requirements. These different types of information are often served by specialized applications. The scale and complexity of applications varies from simple reports to entire decision support systems, depending on the size of the company and the complexity of the business activities being managed. The Web is the common platform for making business-related information available. GIS executive dashboard applications integrate and display data from many sources and use widgets and other controls to interact with and query the underlying data. GIS applications used in ERP make information pertaining to specific management areas more accessible and more easily understood.

## 8.2.3 BUSINESS ANALYSIS

While there is not a clear divide between business operations and business analysis, the business analysis in this section refers to tasks such as evaluating how a retail store is performing; determining who the customers are and where they are located; perceiving what the supply chain looks like; and considering how the store should be laid out, what merchandise should be sold, how much of it should be inventoried, and what types of personnel are necessary to make the store successful. The ability to analyze all these factors is critical for making sound decisions to ensure a business's success in the short term and the long term. Many of these questions have a geographical component and can benefit from the use of GIS technology.

Effective business analysis depends on comprehensive data, and it goes beyond data about a company's products. It requires data about potential customers as well as existing customers, and it requires data that's georeferenced as well as the business data found in tabular spreadsheets. Business analysis typically requires the following types of data:

- Retail information that specifies a business's own products, stores, and customers

- Demographic information such as current population in an area, future population, and the percentages by age, race, education, and gender

- Customer spending that includes categories of purchases and amounts of consumer spending in each category

- Market potential to gauge the demand for goods and services in an area

- Existing business locations such as the locations of major shopping centers, public attractions, and competitors

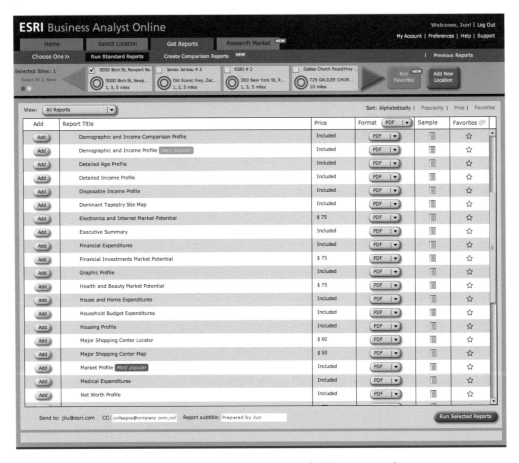

**Figure 8.5**　ESRI BAO uses the SaaS business model to deliver Web GIS as a commodity.

Collecting, managing, and analyzing this data locally can be difficult and expensive for businesses, especially small businesses. However, the emergence of SaaS and the trend of companies to move computer processing from in-house to online or using services "in the cloud" is making these types of analyses more accessible and affordable (see chapter 10). Companies can purchase business analysis data and functions on demand and by subscription. Since a third party owns and manages public cloud services, consumers of these services do not own assets in the cloud but pay for them on a per-use basis (Kouyoumjian 2010). This business model has great appeal because it reduces the complexity of system implementation and eliminates the costs of data acquisition, maintenance, and labor while making information and applications accessible from any desktop, PC, or mobile device.

ESRI BAO is an example of business analysis functions that can be accessed with the use of SaaS and cloud computing technology. BAO, built on top of the ArcGIS Online cloud, combines GIS technology with extensive demographic, consumer spending, and business data to deliver on-demand reports and maps (figure 8.5). Users can select a location on the map and obtain a variety of business reports about the adjacent area, including demographic data (figure 8.6) such as current-year estimates and five-year projections of various population traits; business data such as a list of businesses, major shopping centers, and a retail marketplace profile; traffic counts, which

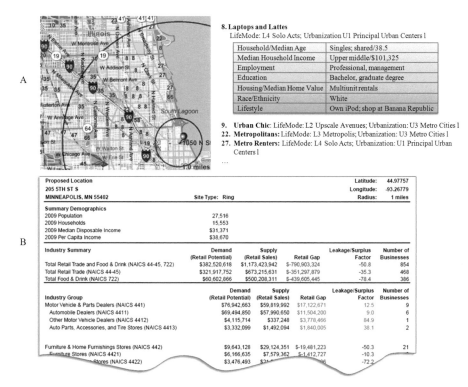

**Figure 8.6** ESRI BAO provides a series of business analysis maps and reports for the locations or areas that users define. Illustrated here are (A) portions of a tapestry segment map and (B) a retail marketplace profile report.

Courtesy of Tele Atlas North America, Inc.; U.S. Census Bureau; and Bureau of Labor Statistics.

describe local traffic patterns; and customer spending and market potential in various categories such as automotive aftermarket, finance, house and home, medical, beauty, recreation, electronics, and restaurants. BAO also provides market segmentation data for the United States. A market segment is a subset of a market and is made up of people who share one or more characteristics that make it likely that they will want similar products or services based on qualities of those products such as price or function. ESRI Tapestry market segmentation, as delivered from BAO, divides U.S. residential neighborhoods into 65 distinctive market segments based on demographic and socioeconomic characteristics (figure 7.6).

Online business analysis products such as BAO deliver commonly used geobusiness analysis functions. They can do the following:

- Perform drive-time, density, or threshold analysis; visualize the results in a map; and make more informed and timely decisions about how to adapt and react to changes in the marketplace

- Analyze trade areas by understanding what businesses currently exist, how far customers have to travel to reach those businesses, and what the potential is to start a business in a particular area

- Evaluate new business sites for growth expansion or examine existing sites to gauge whether the maximum revenue potential is achieved

- Identify the most profitable customers and find more like them by analyzing current spending versus potential spending

- Reach customers effectively through targeted messages and marketing channels by identifying specific market segments and their preferred media and communication choices

Levi Strauss, North America, a division of Levi Strauss & Co. (LS&CO), which manufactures denim jeans, needed a cost-effective solution to manage the approval process for new authorized retailer locations. LS&CO receives numerous retailer application packets each week and needed a solution to streamline its review process. BAO enables LS&CO to see new store prospects geographically in relation to existing stores. If the location of an applicant store is deemed too close to an existing store, the application may not be accepted. LS&CO can further evaluate a site by doing a demographic ring study with one-, three-, and five-mile ring reports. These business, demographic, and consumer household reports help LS&CO to evaluate the business potential of the applicant site and decide whether to accept an application (Levi Strauss & Co. 2007). BAO data and functions are also available via SOAP and REST Web services, as well as by using the Silverlight API. These programming interfaces allow organizations with specialized needs to build custom Web applications that include BAO demographic data reporting capabilities.

Companies with specialized needs can make the geobusiness technology available to multiple users via server-based geobusiness products like ArcGIS Business Analyst Server (BA Server). BA Server (figure 8.7) includes a database containing basemap data, demographic data, customer spending data, and market potential data. This data can be combined with the user organization's own database to support an array of mapping, reporting, and analysis functions. BA Server can perform advanced geographic analysis, such as competitive and site evaluation using simple rings, threshold rings, data-driven rings, drive times, and desire lines (plotted from each customer to the store that serves them). BA Server can identify characteristics for targeting profitable new customers by analyzing the demographic variables using Principal Components Analysis (PCA). With

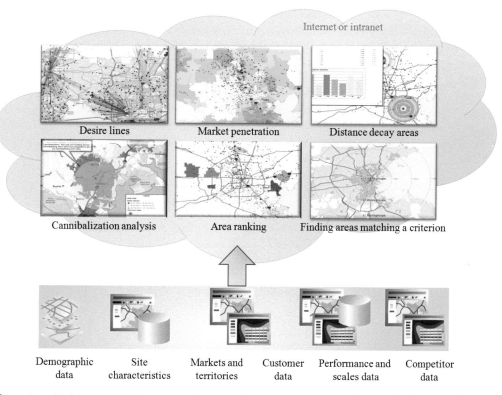

**Figure 8.7** ArcGIS BA Server includes an extensive data package and a series of customizable analysis functions. Organizations can integrate their own business data and build specialized business analysis applications to share with management.

Courtesy of Tele Atlas North America, Inc., and U.S. Census Bureau.

products such as BA Server, business organizations can perform business continuity planning by modeling what-if scenarios for natural or human-made disasters, supplier disruptions, and various transportation options, as well as determine site vulnerabilities based on response times. The analysis results can be presented in map and report templates and shared online, with security controls, with management or across the organization to provide a common platform for informed decision making.

## 8.2.4 LOCATION-BASED SERVICES

Location-based services (LBS) are information services that leverage a mobile user's physical location to provide targeted information (see chapter 5). LBS can be used in geospatially targeted advertising and business operations. Typical examples of LBS include navigation and mapping of points of interest, weather information, location-based advertising, social networking, and tracking a field workforce. LBS applications such as geofencing can help businesses manage assets and workforces.

**Geofencing is an LBS application that sets up a virtual perimeter around a geographic area so that when a mobile device enters or exits the area, a notification is generated.** The notification can contain information about the location of the device and might be sent to a mobile telephone or an e-mail account. This technique has been used to track children and to notify parents when a child leaves a designated area. Law enforcement can also use this technique to keep sex offenders from stalking school areas. From a business standpoint, LBS can improve customer service and optimize workflows. For instance, when field workers such as repair personnel or salespeople enter the "geofencing" around a customer's home, they can automatically be alerted with information updates about the customer or about the services the customer needs to provide more effective customer service.

LBS offers vast opportunities for business applications. Among the predicted top-ten consumer mobile applications for 2012, LBS ranked second because of its high user value in terms of meeting a range of needs, from worker productivity and goal fulfillment to social networking and entertainment (Gardner 2009). Local weather, direction/navigation, and local store locations are among the most popular LBS applications, but mobile users are also interested in getting local alerts and special offers or promotions from nearby stores and other retailers (deGaravilla 2009).

## 8.3 CHALLENGES AND PROSPECTS

While GIS has traditionally been used in government and academia, its applications in business have been on the ascent, especially since the 1990s with the emergence of Web GIS. Many companies use GIS for a variety of applications, including advertising, marketing, retail, logistics, CRM, and ERP. As Web GIS continues to reduce the technical barriers and financial costs of using GIS technology and more business leaders discover the benefits of GIS, businesses will assimilate this technology quickly into both business operations and decision making. Business will continue to be a major area of growth for GIS and Web GIS applications.

### 8.3.1 MOVING PAST THE HURDLES

Technological developments, such as the delivery of GIS as a service, have encouraged the integration of GIS technology into business processes. However, hurdles remain to wider adoption of GIS-based e-business applications—namely, a limited appreciation of the benefits these applications can bring to a business, a shortage of current data on customers, and concerns over security and privacy.

#### LIMITED APPRECIATION OF THE VALUE OF GEOGRAPHY

Although awareness of the value of mapping has greatly increased in the twenty-first century, **many businesses still have a narrow appreciation of the competitive advantages derived from structuring their information and operations along the geographic framework supplied by GIS.** The advent of Web GIS, however, is substantially enhancing the business community's acceptance of GIS. Web GIS is transforming GIS from a technology that is maintained by skilled professionals within a business to a technology that is made available as a commodity, combined with other

IT systems, and delivered over the Web for consumption on demand. This SaaS model is attractive to businesses because it provides the benefits of GIS while eliminating the need for capital investments in hardware, software, geospatial data, and skilled personnel. The task-focused nature of Web GIS applications makes GIS capabilities accessible to a greater variety of personnel within an organization. In addition, as businesses continue moving a greater percentage of their interactions with customers, suppliers, and employees to the Web, they have much greater flexibility in terms of acquiring and delivering products and services. For businesses that have incorporated Web GIS into their workflow, GIS has become one more tool for improving customer interactions.

University business schools are beginning to include GIS courses as part of the curriculum. Some universities, such as Oklahoma State University and the University of Montana at Billings, offer degrees in business geography. However, faculty members in business schools often lack geographic training and may be hesitant to introduce it in their coursework. Moreover, GIS software and data typically are located in the geography department, rather than in business classrooms, which can create an additional obstacle. The strategic value of using GIS in marketing research and analysis, logistics, management science, operations, and information systems is breaking down these barriers. The availability of GIS software site licensing for universities is also playing a role. The Wharton School at the University of Pennsylvania, the world's first collegiate business school, has had a GIS software site license since 2000 and maintains a state-of-the-art GIS lab. The lab is successfully partnering with business to devise innovative solutions using geospatial methods. These real-world demonstrations of the value of GIS to business are key to spreading the appreciation of GIS in business.

## SHORTAGE OF DATA FOR BUSINESS APPLICATIONS

**As with many information systems, GIS business applications require access to comprehensive, accurate, and current data.** "Data quality and the manner in which it translates to quality in the (geographic information) produced becomes a central issue in a business" (Frank and Raubal 2001). Online advertising applications, no matter how sophisticated, rely on current, appropriately scaled basemaps, and large and well-maintained collections of POIs. Business analysis requires demographic and customer spending data at a scale and level of detail that is commensurate with marketing needs.

Detailed demographic data, collected and compiled by the U.S. government, is readily available. Data on age, population, education, income, race, homeownership, and other information is available at the Census Block Group level. The pervasiveness of credit card purchases in many countries makes it possible for companies to obtain information (with identifying characteristics removed) on customer spending habits. Detailed analysis of business customers and markets is relatively easy with this type of information included in the GIS database.

This is not the case in some countries. Detailed demographic data is not always available, or doesn't have adequate spatial resolution to support effective business applications. In predominantly cash economies, customer spending information can be difficult to obtain. Consequently, the potential of using Web GIS applications in business will remain largely untapped for many countries until they can obtain good-quality spatially related data on customers and their spending habits.

## SECURITY REQUIREMENTS AND PROTECTION OF PRIVACY

While the Web has introduced a level of speed, convenience, and customization to business transactions, the downside has been increasing concerns over privacy. Personally identifiable information (PII) can be leaked in many ways. PII is any piece of information that can potentially be used to uniquely identify, contact, or locate a person. In addition to contact and location information, personal information can be unintentionally disclosed through online or network activities. For example, online searches for information on a specific disease might be used as evidence of a health issue. Amassing information relating to personal behavior gathered without the knowledge or consent of users can seriously compromise individual privacy. GIS business applications use sensitive data such as credit card information and can be compromised by hacking or phishing. User names, passwords, or other valuable information can be fraudulently obtained. Business Web applications, including Web GIS applications, therefore, require more robust security to protect businesses' proprietary information as well as safeguard consumers' privacy.

Because online business applications involve sensitive data such as an individual's financial records or confidential business information, there is substantial potential for misuse of the information. With positioning capabilities (GPS based, cellular network based, and the like), mobile phones can leak the owner's location (see chapter 5). While this information, used by Web business applications to deliver targeted contents such as location-based advertising, is generally benign, it can be used for criminal activities that threaten individuals if business systems are compromised or the information is sold without the individual's knowledge or consent.

The misappropriation of sensitive information can lead to identity theft and fraud. A report titled "Identity Theft: Trends and Issues" (Finklea 2009), prepared for Congress by the Congressional Research Service, noted that approximately 9.9 million Americans were reported victims of identity theft in 2008, a 22 percent increase over the previous year. The Federal Trade Commission estimates the cost of identity theft at $50 billion annually. Customers can be discouraged from making online transactions because of legitimate fears over identity theft and fraud stemming from online business transactions and breaches of business databases (Finklea 2009).

Web GIS, however, has also aided in detecting fraud by correlating a customer's home address with the location where purchased items are shipped. This process, called **Address Verification Service (AVS), compares the billing address provided in a transaction with the cardholder's address on file at the credit-card-issuing bank.** AVS returns a code to the payment gateway that indicates verification and accepts or declines the transaction. Data-mining techniques combined with GIS have also been used successfully to quickly identify anomalies in credit card and insurance claim activities and minimize losses.

### 8.3.2 THE FUTURE OF WEB GIS IN E-BUSINESS

Taken as a whole, the benefits of Web GIS for business far outweigh any disadvantages. Web GIS makes Web applications more intuitive and responsive for consumers. For businesses, Web GIS improves the efficiency of operations by integrating business processes and optimizing the use of other information systems. Web GIS makes these business-related tools more widely available as Web applications that do not require extensive knowledge of GIS. It will take a combination of policy making and technology to create a balance between the privacy and security needs of individuals and the needs of business to develop customer products and services with the use of Web GIS.

*"In the broadest sense, technology extends our abilities to change the world; to cut, shape, or put together materials; to move things from one place to another; to reach farther with our hands, voices, and senses."* (Rutherford and Ahlgren 1991)

GIS has greatly expanded the ability of business to reach into existing datasets and acquire detailed information about individual customers. By integrating many existing datasets that use geographic location, businesses can manage and analyze data about consumer behavior on a scale that was previously not possible.

The Web, a pervasive technology originally developed to share scientific work, has generated numerous side effects as it has permeated everyday life. The case is the same for Web GIS. The combination of GIS and the Web has given government, business, and everyday citizens new capabilities. One measure of the impact of Web GIS can be seen in the reported and projected figures for online retail spending. As businesses gain a greater appreciation of the value of GIS delivered over the Web, Web GIS will continue to play a greater role in improving business operations.

## Study questions

1. What is e-business? How did it emerge and develop?

2. What is the value of geography to business? How does Web GIS facilitate geobusiness better than desktop GIS?

3. What type of geospatial data is needed for targeted online advertising, and how is this data obtained?

4. Analyze the characteristics of the business model for Web map-based online advertising.

5. What are CRM and ERP? Give examples of how Web GIS can be used to facilitate them.

6. What geobusiness analysis functions does your, or one of your friend's, business organization need? How can these functions be achieved?

7. Explain Web GIS as a commodity in the SaaS context and give some examples.

8. What factors have slowed the adoption of GIS by business?

9. In what ways do Web GIS applications pose a threat to individuals?

10. How is GIS technology used to detect fraud?

## References

Bambury, Paul. 2006. A taxonomy of Internet commerce. *First Monday,*July. `http://outreach.lib.uic.edu/www/issues/issue3_10/bambury/index.html` (accessed October 26, 2009).

Christensen, Clayton M. 1995. *Disruptive technologies: Catching the wave.* Boston: Harvard Business School Publishing.

deGaravilla, Andy. 2009. Location based services: Why smartphone apps will pay off for advertisers, carriers, application providers. http://blog.compete.com/2009/06/02/location-based-services-applications-carriers-advertisers (accessed March 10, 2010).

Deighton, John, John Quelch, and Hamilton Consultants. 2009. *Economic value of the advertising-supported Internet ecosystem*. Cambridge, Mass.: Interactive Advertising Bureau.

Finklea, Kristin M. 2009. *Identity theft: Trends and issues*. Congressional Research Service.

Frank, Andrew U., and Martin Raubal. 2001. GIS education today: From GI Science to GI engineering. *URISA Journal* 13 (2): 5–10.

Gardner, Dana. 2009. Briefings Direct analysts debate the "imminent death" of enterprise IT as cloud models ascend. *CBS Interactive* http://blogs.zdnet.com/Gardner/?p=3115 (accessed August 11, 2009).

Gerstner, Louis V. 2002. *Who says elephants can't dance? Inside IBM's historic turnaround*. New York: HarperCollins.

Kouyoumjian, Victoria. 2010. The new age of cloud computing and GIS. *ArcWatch*, January. http://www.esri.com/news/arcwatch/0110/feature.html (accessed January 22, 2010).

Levi Strauss & Co. 2007. GIS and Web services help manufacturers find the best retailers. In *GIS best practices: GIS for customer and market analytics*. Redlands, Calif.: ESRI. http://www.esri.com/library/bestpractices/market-analytics.pdf (accessed March 9 2010).

Li, Feng. 2007. *What is e-business? How the Internet transforms organizations*. Malden, Mass.: Blackwell Publishing.

Net Industries. 2010a. Internet Tax Freedom Act. http://ecommerce.hostip.info/pages/629/Internet-Tax-Freedom-Act.html#ixzz0VC1XMSSG (accessed October 27, 2009).

———. 2010b. IBM Corp.: Shift to e-business services. http://ecommerce.hostip.info/pages/561/IBM-Corp-SHIFT-E-BUSINESS-SERVICES.html (accessed October 27, 2009).

Onsrud, Harlan J., Jeff P. Johnson, and Xavier Lopez. 1994. Protecting personal privacy in using geographic information systems. *Photogrammetric Engineering and Remote Sensing* 60 (9):1083–95.

Pick, James B. 2008. *Geo-Business: GIS in the digital organization*. Hoboken, N.J.: John Wiley & Sons.

Pratt, Monica. 2005. Mapping better business strategies: Integrating GIS and business intelligence. *ArcUser*, July–September.

Reynaert, Bart, Patrick Vercruyssen, and Rene Horemans. 2005. GIS and beyond: Integration with SAP improves business processes. *ArcUser*, July–September.

Rutherford, F. James, and Andrew Ahlgren. 1991. *Science for all Americans*. Oxford: Oxford University Press.

Symonds, Matthew. 1999. The net imperative: A survey of business and the Internet. *The Economist*, no. 26.

Tai, Rob. 2009. Measuring the impact of the Internet on the economy. Google Public Policy Blog, June 10. http://googlepublicpolicy.blogspot.com/2009/06/measuring-impact-of-internet-on-economy.html (accessed October 29, 2009).

Yi, Xing, Hema Raghavan, and Chris Leggetter. 2009. Discovering users' specific Geo intention in Web search. Proceedings of WWW 2009 Madrid. http://www2009.eprints.org/49/1/p481.pdf (accessed October 30, 2009).

# WEB GIS APPLICATIONS IN E-GOVERNMENT

Since the inception of GIS, agencies are its main users. The first GIS framework was invented in the 1960s in response to the Canadian government's need to inventory and manage land and natural resources. Since then, governments have adopted GIS methodology, first at the project level, then at the departmental level, and later as an enterprise-wide information system. Since Web GIS came to life in 1993, governments have quickly adopted it as one of the main information technologies (IT) to implement and enhance e-government. They have used IT, principally the Web, to improve the level and quality of government services. Now, government agencies routinely employ Web GIS to deliver services to citizens, to streamline government operations, and to coordinate activities among agencies.

The first section of this chapter describes how e-government developed, and the second section provides examples of the ways in which GIS, and Web GIS, help government to function more effectively and efficiently. It is worthwhile to note that Web GIS applications in e-government are also covered by the examples and case studies in other chapters of this book. The final section explores the challenges and prospects facing e-government.

# 9.1 INTRODUCTION TO E-GOVERNMENT AND GEOGOVERNMENT

Rapid advances in IT, especially the explosion of the Web, coupled with increased global competition, have driven the evolution of government. Most countries in the world today are looking to adopt e-government, which promises to improve government efficiency, reduce costs, increase transparency, and strengthen citizen participation. Because geography is an important aspect of government services, GIS is a critical component of government IT infrastructure and plays an important role in the digital evolution of e-government, leading to increased use of geogovernment.

## 9.1.1 THE EMERGENCE AND DEVELOPMENT OF E-GOVERNMENT

Whether it is called e-government, e-gov, digital government, or online government, **e-government is the use of IT, principally the Web, to improve the level and quality of government services.** The Web has become a tool for collaboration, communication, and service in four areas:

1. **Interactions within government**
2. **Interactions between governments**
3. **Interactions between government and business**
4. **Interactions between government and citizens**

Decades before the advent of the Web, governments embraced the concept of applying IT to improve internal processes. The huge and rapidly increasing volume of data maintained by federal agencies such as the U.S. Social Security Administration (SSA), U.S. Postal Service (USPS), and U.S. Census Bureau led to the government adopting computers to automate processes and to increase the accessibility of information—within government departments, between government agencies, in government relationships with business, and between the government and public citizens.

Perhaps the most visible legacy of e-government is how the Web has changed the ways in which government interacts with citizens. In a precursor to government on the Web, the Clinton administration in the early 1990s saw computers and telecommunications as the way to communicate with the public. President Clinton created the National Performance Review, later renamed the National Partnership for Reinventing Government, which aimed to create government that was more responsive and more effective. This legislation put forward the concept of reinventing, or reforming, the government through information technology. It underscored that the use of the Internet and of Web sites was best practice for communicating with staff and citizens.

The motivation for government Web sites was the assumption that the Web would be readily accessible to the majority of citizens. By the end of the 1990s, the number of U.S. households with Internet access had increased to nearly 50 percent. A comparable number of people had Internet access through the workplace. The Clinton administration saw the Web as a tool to make government more accessible, responsive, and transparent by supplying universal access to information and by allowing citizens to carry out transactions more quickly, more conveniently, and more cost-effectively.

As a result of the Clinton-era reform, by the end of the 1990s, federal and state government agencies, and most local government agencies, provided Web-based services. With more and more government agencies providing Web-based services, citizens needed a means to quickly discover the services they were seeking. To meet this need, in September 2000, the U.S. government launched the FirstGov.gov Web site, later changed to usa.gov, as the official Web portal of the U.S. government. It provided a single point of entry to every federal agency and to state, local, and tribal governments. As such, it is the most comprehensive site in, and about, the U.S. government. Initially, FirstGov.gov provided access to 47 million U.S. government Web pages. The FirstGov portal was clearly designed for broader public use since it provided the ability to search by topic rather than by agency allowing citizens to more easily discover desired information. Privacy standards were also enacted to safeguard citizens' communications with the government. The E-government Act of 2002, under President George W. Bush, continued the strategy of using the Web as the medium of communication with the public. The mandate for government agencies to use the Web to communicate with the public was even more explicit. The act promoted "the use by the government of Web-based Internet applications and other information technologies, combined with processes that implement these technologies to (A) enhance the access to and delivery of government information and services to the public, other agencies, and other government entities, or (B) bring about improvements in government to operations that may include effectiveness, efficiency, service quality, or transformation" (U.S. 107th Congress 2002).

The IT research firm Gartner categorized these cross-level functions in a 2002 report on government trends that focused on administration and finance, transportation, public safety, human services, health, criminal justice, natural resources and the environment, and public works. The report noted that federal, state, and local governments share similar concerns and perform complementary tasks, but their style of interactions with the public and with other agencies may vary and is reflected in the type of Web site each agency creates.

The role of state governments as an intermediary between the federal government and local governments has made state governments the natural hub for integrating government operations and exchanging information. States tend to have more direct transactions with citizens than federal agencies do, and many of these transactions can be easily performed on the Web. State agencies issue driver's licenses, fishing licenses, and car registrations, all transactions that are amenable to Web GIS. The e-commerce style of applications used for these transactions is familiar and generally well accepted by the public, and the cost savings to state agencies has encouraged many states to make the transition to using online applications via e-government.

Local government, for many people, is often the face of government. If people want a pothole filled, want to know where to vote, or want information on recycling, they turn to local government for help. Local governments were quick to adopt the Web as a way to provide centralized access to information on government services. Web sites are also a way for elected officials to communicate directly with citizens, solicit support for policies and programs, and garner feedback. Local government Web sites also serve as a conduit for business proposals between government and the private sector. Government Web sites announce projects and contract offers and allow companies to file their bids online, streamlining the interactions of the private sector with local government agencies.

The adoption of new Web technologies, however, has not been universal for government agencies. Government Web sites continue to retain a mix of static information pages, along with database-driven interactive pages and pages with links to downloadable forms. Still, most government agencies have moved to restructure their Web sites to better serve the public, by automating transactions and being more responsive. They have found that service-oriented Web sites can actually save time and money over having staff members constantly update static Web pages or answer the same questions from the public over and over. Government Web sites free government staff to work on other tasks and also supply information that can be used for internal processes. On the public service side, the Web provides government information twenty-four hours a day, seven days a week, without requiring citizens to pay a visit to a government office. The Web also provides a more efficient way to deliver federal services such as food stamps, Social Security benefits, and Medicaid. One area in need of further development in terms of Web access is the ability of government to provide direct interaction with individual citizens. Still, government services have improved as a result of the government using the Web to communicate with the public via e-government.

In Web 2.0, the Web is no longer just about passively reading information or watching events. Instead, it is all about sharing, socializing, and collaborating on the Web. The term "government 2.0" was coined to refer to attempts to apply the social networking and integration advantages of Web 2.0 to government practices (Eggers 2005). President Obama, whose presidential campaign was heavily influenced by Web 2.0 technology, emphasized that the government should be transparent, participatory, and collaborative in one of his first memorandums in 2009 (Obama). An important part of achieving these goals includes effectively using Web 2.0 technologies such as blogs, wikis, social networks, Web services, and mashups to communicate government information to government employees and to the public.

## 9.1.2 THE VALUE OF GIS TO GOVERNMENT

Approximately 80 percent of government records include a geographic component that ties that record to the landscape, whether it's a street address, parcel number, or ZIP Code. GIS technology makes use of this component to help government agencies see spatial relationships that aren't as readily apparent from simply viewing the data in a table or a report. Being able to analyze data more effectively, and see more clearly what it means, brings an additional dimension to decision making that's made possible with GIS.

GIS technology was originally developed to help governments manage land. Roger Tomlinson is credited with creating the first GIS as a tool for performing the Canada Land Inventory in the 1960s. Governments, at all levels, have continued to make extensive and innovative use of GIS methodology. The United States has adopted GIS software as the central technology for Homeland Security, for example, because it allows agencies across government to share authoritative data and a common operating view. State transportation departments use GIS to construct, inventory, and manage thousands of miles of state highways. Local planning departments use GIS to integrate information on demographics, infrastructure, and natural resources to model scenarios that suit public needs.

Government agencies use the geographic framework supplied by GIS technology to organize, analyze, and visualize all types of data to solve problems, make policies, and improve government processes. As Charles G. Groat, director of the U.S. Geological Survey (USGS) from 1998 to 2005, noted, "Doing good science is not enough. It has to be effective in the community." For science to play an effective role in a rapidly changing world, Groat stressed, decision making can be enhanced by integrating what is learned through science with other types of knowledge. The number and currency of data sources that can be accessed and considered in decision making has greatly expanded with GIS on the Web (Pratt 2000).

Initially, the use of GIS technology—like e-government—was pursued as a way to streamline government processes and to improve government services, but its scope was limited to a single project, or at most, to a single government department. It was adopted as a tool for meeting a specific goal or accomplishing a certain task. Once the GIS concept was introduced as a way to look at government data and processes, it was applied to a number of government activities, from asset management for utilities to all facets of planning to monitoring police and code enforcement activities and building inspection to decision support at all levels.

The Planning Department of the City of Ontario, California, for example, first used GIS technology in the late 1980s to compile a land-use database from various sources. A GIS department was set up to centralize access to the technology, and the use of GIS soon spread to the police, building, fire, finance, and other departments. The use of GIS has become a true enterprise implementation over the course of three decades, improving many of the city's business processes. The city's Revenue Division realized in the early 1990s that noncompliance with local business licenses was costing the city thousands of dollars in lost fees, lost sales tax, and uncollected penalties. Instead of physically inspecting all the city's businesses, the Revenue Division used GIS to find offending businesses without having to leave city hall. All parcels with commercial or industrial uses, such as shopping centers or industrial parks, were queried and cross-referenced with the business license file to identify parcels that lacked licenses. Using the owner addresses attached to the parcel record, the Revenue Division mailed letters to parcel owners requesting a list of tenants. The division then ran the tenant lists against the business license file to identify businesses operating without a license and notified the owners how much they owed in fees, taxes, and fines. With the use of GIS, city staff had to scrutinize businesses just once to generate more than $100,000 in business revenues.

An organization can get the best returns from GIS technology by using a broad implementation that reuses data in multiple applications, spreading the costs of hardware, software, and maintenance across multiple department budgets. The overall quality of operations, as exemplified by the City of Ontario, benefits from the use of GIS making government processes "faster, cheaper, and better." GIS technology makes government faster by automating processes and communicating information more quickly, improving government response times. GIS makes government cheaper by coordinating operations across the organization and saving staff time and expense. GIS helps government deliver a better level of service by decreasing response times, enabling more intelligent management of resources, and increasing the efficiency of operations.

The same improvements in technology that have enhanced mainstream IT also enhance GIS technology, including faster processing, increased bandwidth, and greater storage capacity. Mobile technologies and real-time networks have also extended the reach of GIS by bringing the technology into the field and using the most current data. As GIS software has evolved, it has moved across many platforms. From its beginning on mainframe computers, GIS has moved to minicomputers, to workstations and PCs, and now to the Web. The amount of geospatial information being collected and maintained has increased so rapidly that designers of Web GIS applications and systems have gone from making do with scarce data resources to effectively integrating what often seems like an overwhelming amount of data. By integrating GIS with other IT systems, Web GIS developers have made the application of GIS technology to management, analysis, visualization, and decision making more feasible and more attractive. The ability of government decision makers to consider large volumes of data in a timely fashion has expanded government's use of GIS applications to a level that was previously not possible.

The use of GIS technology has become so integrated with daily operations that some government organizations could not perform at the same level without GIS technology. The advent of GIS in government has been compared with the introduction of the personal computer for its ability to make its users more productive. Before the introduction of GIS technology, city planning staff could plan, maybe, one or two proposals in a given time frame. Elected officials had to make a decision based on those limited choices. With the use of GIS, the same staff can present a variety of well-researched proposals to decision makers. GIS technology gives organizations the ability to deliver services that were not previously feasible.

In the early 1990s, governments were squeezed by two opposing forces—limited budgets and citizens' high expectations of government services. Fewer dollars meant fewer staff members, but those staff members were still expected to perform many tasks. The introduction of GIS technology ameliorated the situation and greatly improved the effectiveness and efficiency of government, especially as GIS moved from being a tool for a specific project to becoming an information framework for an entire department.

For years, GIS remained primarily a tool used by GIS professionals, who performed all the GIS tasks for an organization. This workflow caused a bottleneck, however, that limited the full potential of using GIS technology. Meanwhile, the growth of the Web was changing the public's expectations of what was possible online, and citizens were demanding better access to government information and services. The types of information that individuals could access online was increasing exponentially, and the information was available at a moment's notice. The public began to expect the same sort of instant gratification from the delivery of government services. "One stop" and "self-help" became the bywords for this model of service delivery.

Since location is vital to identifying much of the information the public typically seeks from government agencies, the role of GIS in e-government has continued to grow. Many government Web sites adopted Web GIS to deliver city planning and zoning maps, parcel and permit inquiries, information on everything from earthquakes to diseases, and locations of registered sex offenders. While most Web GIS applications in e-government feature a top-down information flow, there are increasingly more government Web sites that use Web 2.0 principles and technologies, including social networking. Echoing the idea of government 2.0, these new Web GIS applications prompt public participation and are often used as a means to collect georeferenced data—that is, inviting public comments on city planning, collecting the latest information on community events, and inviting ideas about how to improve local traffic management. As government Web sites have evolved, the Web has made the interactions between citizens and government agencies more convenient, more accessible, and more efficient. The adoption of Web GIS has greatly enhanced this process, with a simple user interface that requires no special knowledge to use and that is accessible anywhere the Internet is available. Web GIS allows citizens to interact with the government without ever leaving their homes (figure 9.1).

**A**

**B**

**C**

**Figure 9.1**　Traditionally, citizens have (A) traveled to city hall to learn about city planning and zoning decisions and offer opinions at public hearings. With the aid of Web GIS, citizens can get government information and provide feedback from anywhere, including (B) from the comfort of home and (C) out in the field.

Web GIS has served to streamline government, both in internal operations and in collaborations across the organization. Indeed, the reach of Web GIS is growing toward the GeoWeb, a large, distributed collaboration of knowledge and discovery that promotes and sustains worldwide sharing and interoperability (Pratt 2006). Instead of providing wide access to a single source of data, the GeoWeb can bring together vast stores of dynamic and disparate data, scattered across different agencies or served from the commercial cloud or the government cloud. Web GIS can present hidden patterns in an easy-to-understand way and supply mapping and analysis capabilities through a simple Web browser interface that allows the public to benefit from geospatial tools and data. Many kinds of clients—smart clients, Web clients, and mobile clients—can consume the data, information, and visualizations that are produced through Web GIS. These interactions form a technical basis for government operations and facilitate GIS-based collaboration, that is, geocollaboration (see chapter 10), across multiple government agencies.

## 9.2 TYPES OF APPLICATIONS

Many Web GIS applications have been developed for e-government. This section introduces three types of e-government applications. While there is not a clear division among these categories (for example, a Web site that provides information to the public can also be part of a government application), the first type is designed to provide information to the public in a timely and cost-effective manner. The second type of e-government application allows for two-way communications that serve to make government more responsive to the public. The final category encompasses e-government applications that improve government operations and enhance decision support.

### 9.2.1 PUBLIC INFORMATION SERVICES

Initially, government Web sites adopted the same approach to supplying information that early business sites did: the Web site as a billboard. Communication was one-way, and Web sites were organized around meeting the needs of a department rather than providing the types of information a visitor might be seeking. However narrowly focused these Web sites were, the public still found them useful, and they were popular. These Web sites contained the kind of information (e.g., hours of operation, schedules of city services, notifications of public hearings) that previously were available only by calling or visiting a government office.

Providing an intuitive interface that the public could use to access other types of government information proved to be more challenging. Since most of the information that governments deal with has a strong connection to place—for example, property taxes are assessed by parcel, garbage trucks are routed on city streets, polling place locations are determined by the home address of voters—governments turned to maps to provide the public access to these types of geographic information. Governments combined the Web with the relatively new technology that had been used to manage information by its geographic location, GIS, to organize this information and deliver it to the public. Thus, governments began using Web GIS to supply information to the public on property taxes, garbage pickup days, and polling places, easily and intuitively. As Web technologies evolved, communications between government and citizens became more customized to the visitors to a Web site. Rather than wading through all the Web pages on a given topic, visitors to a government Web site could query just the information they wanted, based on a specific location, status, or date parameter, for example.

However, the public was not the original target audience for government Web sites that used GIS technology. Government agencies realized that all they needed was a simple Web browser to extend the benefits of GIS datasets and applications. Staff members who were not GIS specialists could use GIS technology to do daily tasks. GIS-based self-serve applications were designed for government staff working at service counters in government offices. With these self-serve applications, staff members could use geography quickly to locate the information they needed to perform daily tasks and to answer questions from the public.

Because providing information to the public is such a fundamental task of government, these applications were moved from the intranet to the Internet to make information more available to the public. The popularity of these applications led government departments to expand their use to deliver all kinds of information. The following agencies use Web GIS to provide information on a daily basis:

- **The U.S. Geological Survey** delivers near-real-time information on natural disasters, including earthquakes, floods, hurricanes, landslides, tsunamis, volcanoes, and wildfires (see figure 1.11). This information is used to keep the public informed as well as to assist agencies responding to these events.

- **The National Weather Service** publishes information on real-time weather conditions, as well as climate and weather forecasts, which are used by the public, the transportation and shipping industries, and emergency responders, among others.

- **The Centers for Disease Control and Prevention (CDC),** part of the U.S. Department of Health and Human Services, uses GIS to make the information it collects on infectious disease outbreaks, environmental hazards, chronic health conditions, and the status of the population's health accessible and comprehensible.

- **The Federal Emergency Management Agency (FEMA)** creates and maintains flood maps to let individuals and insurance companies evaluate risks.

- **The U.S. Environmental Protection Agency (EPA),** via the My Environment Web site, delivers information on local air and water quality. The EPA uses GIS technology to incorporate information from federal, state, and local government sources as well as private companies. Visitors can search the site by ZIP Code or address.

- **Local governments** use parcel boundary data as a framework to give the public easy access to information on real property ownership, planning, and zoning.

## LOCAL GOVERNMENT APPLICATIONS

Web maps are often used at the local government level to deliver local information on zoning, property assessments, crimes, underground pipe repairs, and street closures. The City of Sacramento, for example, began incorporating GIS on its Web site in 1998. Although the city initially used GIS on its Web site to help planning department staff answer public inquiries, the public was soon getting answers for itself directly from the Web site. The city Web site offered three kinds of maps: general, property, and major crimes. General maps supplied information on council districts, neighborhood service areas, and city services. Property maps furnished information from the assessor's office, as well as on zoning and the locations of public schools, parks, streets, and rivers. Major crime maps show the locations of kidnappings, assaults, thefts, burglaries, and homicides.

Web GIS caused a powerful shift in the way people used the Web. Instead of relying on someone else to interpret the data for them, users could perform their own queries and analysis to come to their own conclusions about what the data revealed. In just three short years, the number of visitors to the Sacramento Web site tripled. Web GIS proved to be a powerful way to deliver government services more cost-effictively. The city estimated, in 2001, that without the site, more than 500 people a day would have had to visit or call city hall to get answers to their questions. By calculating the time it would take staff members, multiplied by their hourly rate, to answer those questions, the city estimated that it saved $1.8 million in a year by answering the needs of citizens with the use of Web GIS (Walters 1999).

In another example of providing better government service, the City of Greeley, Colorado, created rich, intuitive Web applications to allow the general public to quickly search for information. The city's ORIGIN Property Information Map (figure 9.2) was developed to address the needs of the Community Development Department, the Planning Division, and the general public. City planners were spending approximately three to four hours a day on the phone answering questions from the community on general property information. Since 2009, the ORIGIN Property Information Map has given the public direct access to parcel data without having to call a city planner. Similar to the case of Sacramento, which uses Web GIS as a means to provide information to citizens, giving them greater flexibility and convenience to explore city data, the City of Greeley

**Figure 9.2** The City of Greeley, Colorado, makes extensive use of Web GIS applications to deliver services to residents. This application gives instant access to a wealth of information about any parcel in the city—from zoning to school districts to the location of the nearest fire hydrant.

Courtesy of City of Greeley.

has also released city employees to perform other duties. Web GIS enables cities like Sacramento and Greeley to provide better citizen service while reducing government costs.

## MEGAN'S LAW

Web GIS supports the implementation of Megan's Law. This law requires persons convicted of juvenile sex crimes to notify local law enforcement of their current address and employment. Law enforcement must then make information about local registered sex offenders available to the public. Although there are different methods for making the required disclosures, such as through newspapers, pamphlets, and public Web sites, many state and local governments have adopted Web mapping as the most intuitive way for people to process the information. For example, the California Department of Justice Internet Web site (`http://www.meganslaw.ca.gov`) lists designated registered sex offenders in the state. It enables citizens to search for sex offenders by county, city, or ZIP Code or by proximity to an address, school, or park. It displays the results on a map and provides detailed personal profile information such as the offender's name, picture, appearance, characteristics, address, and criminal offenses.

## CENSUS DATA

State and federal agencies also need to find an efficient and cost-effective way to provide information to citizens. These agencies selected the Web as the means to share information, and they adopted GIS technology, which had been widely applied to many activities of state and federal government, to better convey that information.

The U.S. Census Bureau, which exemplifies the integration of GIS with business workflows, uses GIS technology to gather, process, and disseminate statistical information about the United States, and then uses Web GIS to share its work with other government agencies and the public. After computerizing the statistical aspects of the census, the Census Bureau implemented a program in the 1960s to compute the geography of the census, GBF/DIME (Geographic Base File/Dual Independent Map Encoding). GBF/DIME, which codes the topology of street segments, revolutionized the census, and it also revolutionized GIS in general. In the early 1980s, GBF/DIME was succeeded by the TIGER (Topologically Integrated Geographic Encoding and Referencing) system. The TIGER system, created in collaboration with USGS for the 1990 decennial census, enabled the Census Bureau to associate the household data it collected by physical address with census units and political boundaries. TIGER/Line files, which comprise a digital database of the streets, rivers, legal boundaries, census statistical boundaries, and other geographic features for the entire United States, can only be visualized with the use of GIS software.

In 1994, the Census Bureau became one of the first government agencies to establish a Web portal, `www.census.gov`, which gave citizens, businesses, and other government agencies direct access to a storehouse of bureau statistics. American FactFinder, created in 1998, let visitors to the Web site generate maps and tables that incorporate multiple datasets (figure 9.3). The State and County QuickFacts Web site thematically maps a summary of business statistics at the state and county levels. The maps generated by both sites let visitors quickly grasp the meaning of statistics displayed in census tables. The Web sites support numerous visits daily and have helped the bureau reduce the number of paper documents it generates. According to Louis Kincannon, former director of the Census Bureau, "Ten years ago, the 1990 census results took up more than 450,000 pieces of paper which, when stretched out, would have spanned 78 miles. Posting Census 2000 results directly on the Internet has enabled us to eliminate more than 75 percent of our paper products" (U.S. Census Bureau 2004).

## TRANSPARENT GOVERNMENT

The use of GIS technology to organize and manage data and to make complex information more comprehensible and accessible introduced a new degree of transparency and accountability for government. Having information on how government runs engages citizens more effectively in government processes, and GIS technology can parse that information into more coherent forms. Traditionally, governments have been very good at tracking the funds spent on government programs but have performed less well in monitoring the outcomes of these programs.

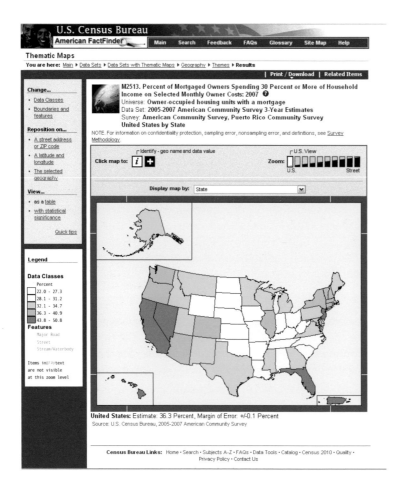

**Figure 9.3** The U.S. Census Bureau's American FactFinder site makes demographic information easy to access and comprehend. Dark green areas on the map indicate areas where homeowners spend 30 percent or more of their monthly household income on their mortgage payment.

Courtesy of U.S. Census Bureau.

When Martin O'Malley was the mayor of Baltimore, he created an expanded version of a GIS-based program developed by the New York City Police Department to reduce crime. Called ComStat, the Baltimore program identified high-crime locations so that police resources could be concentrated in those areas. In 2000, O'Malley began using GIS across city departments via CitiStat as a tool for better government. CitiStat initially focused on cracking down on the rampant absenteeism in the city's workforce. After saving the city $13.2 million in its first year, the program was expanded. The city now uses it to manage all city programs, from potholes to parks, saving $100 million in its first four years. The success of CitiStat, which earned the Innovations in Government award from the Harvard University Kennedy School of Government in 2004 (City of Baltimore 2004), has led eleven other cities, including Washington, D.C., to adopt this approach.

As mayor, O'Malley endorsed GIS as an important part of a performance-based approach to government. "By analyzing our performance in a geographic context, we are able to reduce operating costs, increase revenue streams, and improve the quality of service we deliver to citizens. An efficient government is one that uses resources responsibly and effectively, and this approach to operations helps us achieve that," O'Malley said (*ArcNews* 2009).

When O'Malley was elected governor in 2006, he continued using GIS as a tool for improving government with StateStat. With StateStat, O'Malley pioneered the extensive use of GIS as a tool for measuring state performance and monitoring government accountability. The State of Maryland StateStat application uses the Web as a delivery system to make information on the performance of government departments and programs accessible to Maryland's citizens and government agencies via Web GIS.

Similar to CitiStat, StateStat monitors the performance of state agencies to make them more efficient and accountable by publishing geographic data online for public access. Beginning with public safety, health care, and social services, StateStat now encompasses the Maryland State Police and the Departments of Agriculture, Business and Economic Development, Environment, General Services, Health and Mental Hygiene, Housing and Community Development, Human Resources, Juvenile Services, Labor, Licensing, and Regulation, Natural Resources, Planning, Public Safety and Correctional Services, and Transportation. StateStat gives citizens access to timely reports from all participating departments on how their tax dollars are being spent. It provides contact information to provide for public feedback and to allow citizens to make requests for state services.

The state has two additional oversight programs, BayStat and GreenPrint, which make use of Web GIS to keep the public informed. BayStat tracks efforts to rehabilitate Chesapeake Bay, integrating data from the Departments of Agriculture, Environment, Natural Resources, and Planning. GreenPrint monitors conservation efforts. It is the first Web-enabled map in the United States to show the relative ecological importance of each land parcel in the state. GreenPrint has a dual mission: it lets citizens learn what has been done to protect the state's lands; and it creates public consensus by raising awareness of the ecology of the state's forests, wetlands, and sensitive habitat.

The federal government also uses Web GIS to provide accountability for government agencies and to ensure government transparency. Obama has made the Web the primary conduit for supplying information through the American Recovery and Reinvestment Act (ARRA) of 2009. The Recovery.gov Web site (figure 9.4), launched in September 2009, provides access to up-to-date information on how $787 billion in economic stimulus funds are being spent. Web GIS functionality gives citizens a way to provide feedback on the effectiveness of stimulus spending.

In response to historically high unemployment rates as a result of the recession that began in December 2007 and a continuing decline in the gross domestic product, ARRA provided funding for hundreds of projects with the primary goal of jump-starting the national economy and creating jobs. ARRA stressed transparency in government by facilitating public tracking of stimulus spending. Numerous Web GIS mashup sites have been created to allow tracking of stimulus spending.

The State of Oregon, for example, launched a site in March 2009 to track stimulus spending in the state. Built with the ArcGIS API for Flex, the Oregon Stimulus Tracker maps the location and progress of ARRA projects by using reports from the agencies that administer the funding. Each month, agencies update this information directly through the Web GIS tracker application. The state then takes the unemployment data it collects and matches the occupation codes of the surrounding labor pool with the skills needed for local projects, providing stimulus jobs for people out of work.

**Figure 9.4**   The Recovery.gov Web site was launched by President Obama to communicate with the electorate about the federal government's economic stimulus package.

Courtesy of Recovery.gov.

## KEY CONSIDERATIONS

To provide information to the public, government agencies must make key considerations regarding timeliness and access. Among these are the following:

- **Does the information pertain to citizens' needs?**

- **Does it have enough coverage and depth to meet users' needs?**

- **Is the information provided at the right time?** While historical data is useful for some analyses, the most current information about unfolding events is needed to provide early warnings and facilitate emergency response.

- **Is the information provided via the appropriate method?** Geographic information can be delivered as text, static maps, interactive maps, Web feeds, or Web services. For example, USGS delivers earthquake information as text, static maps, and also interactive maps to let citizens closely examine where earthquakes occur. The earthquake information is also delivered as GeoRSS and KML feeds, which are essentially Web services that can be viewed using a variety of geobrowsers and integrated with other applications, including the ESRI Situational Awareness Bundle (`http://www.esri.com/software/arcgis/situational-awareness/`), the Natural Hazards and Vulnerabilities Atlas (`http://www.pdc.org/atlas`) by the Pacific Disaster Center, and the Global Disaster Alert and Coordination System (`http://www.gdacs.org`) by the European Commission Joint Research Center.

- **Does the application provide a good user experience?** Government Web GIS applications should be easy to use and easy to understand to effectively convey the message and engage the public.

## 9.2.2 TWO-WAY INTERACTIVE COMMUNICATION

In addition to providing public information services, which were often one-way communication from government to citizens, many government Web sites have extended government's reach from merely disseminating information to collecting information as well. At its most basic level, this move toward two-way communications consisted of providing contact e-mail addresses for a specific department or public official or providing links to forms that citizens could download, print, and return to a government agency. Although the communications remained primarily unidirectional, interaction between the government and citizens was possible. Still, there were typically significant delays in the government's response to issues a citizen might raise in a query.

As Web technologies evolved, government sites also began allowing visitors to post comments, make service requests, and perform many types of online transactions, including filing tax returns, applying for grants, and filing building permits. In the Web 2.0 era, user involvement is a defining characteristic of the Web. A variety of different types of applications such as wikis, blogging, and social networking have emerged and become popular, resulting in huge volumes of user-generated content (UGC). These applications reveal new approaches for spurring public participation in government Web sites.

The capabilities of Web GIS applications have transformed one-way communication into a platform for supporting citizen participation in government. The rapid growth of volunteered geographic information (VGI), part of the Web phenomenon of UGC facilitated by Web GIS, engages private individuals in updating and supplementing more traditional geographic data-collecting methods (Goodchild 2007) and is employed in a number of pioneer Web applications to harness citizen contributions (see chapter 10.1).

### PUBLIC HEALTH

Public health threats are an issue that brings government agencies together through the use of Web GIS. Since early 2000, for example, the use of Web GIS across federal, state, and local lines has improved the government response to outbreaks of West Nile virus (WNV). WNV spread into North America in 1999, with the first human cases usually appearing within three months of the first appearance of infected birds in an area. Mosquitoes that bite the infected birds spread WNV to animals and humans.

The use of Web GIS technology has allowed various levels of government to effectively map and track the spread of WNV. In California, the state Departments of Public Health, and Food

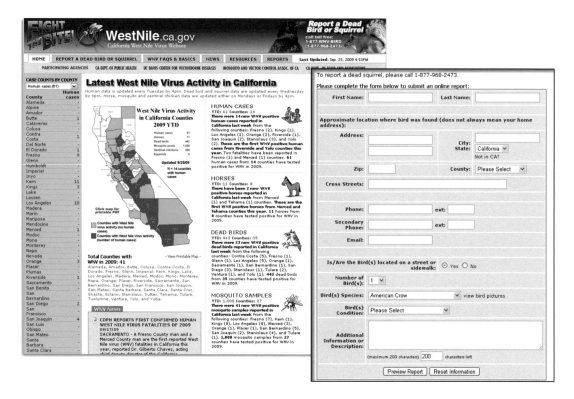

**Figure 9.5** WestNile.ca.gov coordinates the sharing of information across state agencies and supports the collection of information from private citizens by letting them report dead birds and squirrels that might spread the virus.

Mosquito logo designed by Gary Sky, illustrated by Vicki Gullickson. © www.FightTheBiteColorado.com. Used with permission. Courtesy of California Department of Public Health.

and Agriculture, in cooperation with the University of California, Davis, Center for Vector-borne Diseases and the Mosquito and Vector Control Association of California, support a Web site that reports human cases, as well as diagnosed horses, dead bird counts, and positive mosquito samples to keep the public and affected agencies informed (figure 9.5). The site also lets staff and citizens report dead birds or squirrels, which may carry the disease. The collected information on suspected cases is geocoded, and then correlated and aggregated to help public health officials coordinate countermeasures to protect the public.

## LAND-USE PLANNING

Web GIS provides a mechanism for directly involving citizens in government processes without requiring citizens to be physically present at meetings or hearings. Citizens can submit their georeferenced opinions on land use by marking and annotating maps available on government Web sites (figure 9.6). The E-Gov for Planning and NEPA (National Environmental Policy Act) project, or ePlanning, for example, is an early Web application from 2002 that was developed so the

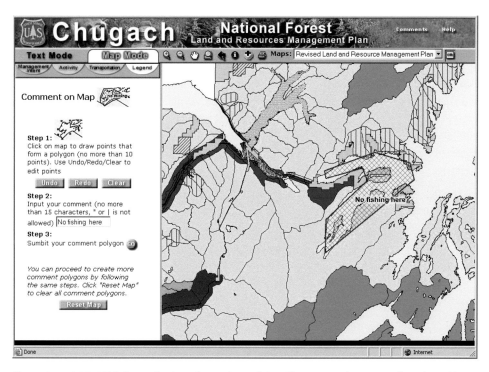

**Figure 9.6** A 2003 U.S. Forest Service pilot study used the ePlanning application to allow the public to mark up and annotate planning maps for the Chugach National Forest in Alaska and to submit georeferenced comments regarding Forest Service proposals.

Courtesy of U.S. Forest Service and Bureau of Land Management.

federal government could capture public input on the disposition of public lands. The ePlanning Web site helped coordinate the planning process across the Bureau of Land Management (BLM) and the U.S. Forest Service (USFS) by involving citizens more directly in the planning process (Zulick 2004). BLM administers 261 million acres of public lands, and USFS administers 193 million acres of national forests and national grasslands. The BLM and USFS use Web GIS to carry out the mandate of the National Environmental Policy Act of 1969 "to declare a national policy which will encourage productive and enjoyable harmony between man and his environment; to promote efforts which will prevent or eliminate damage to the environment and biosphere and stimulate the health and welfare of man; to enrich the understanding of the ecological systems and natural resources important to the nation; and to establish a Council on Environmental Quality"(CEQ NEPAnet 1970).

NEPA fosters an approach to land-use management that makes planning information accessible to the public and requests public comment during a specified time period. The agency then evaluates and responds to these land-use comments. Because geography is central to land-use planning, NEPA used Web GIS technology as the means of delivering information and collecting public comment. The ePlanning site provides a geographic framework for locating and accessing land-use planning and NEPA documents. At its inception, the project's most novel aspect was its use of Web GIS as a mechanism for participatory, collaborative, and community-based land-use planning. Both government staff and members of the public can use Web GIS technology to read agency documents on current planning projects, view maps related to the plans, and submit comments on issues such as recreation, wildlife, and energy. Interactive documents link specific sections of the text to specific features on the map, and users can click map features to view relevant text and to make online comments. The ePlanning application facilitates the government-to-government and government-to-citizen exchange of information and ideas through the use of Web GIS technology.

## TRANSPORTATION

Different types of Web 2.0 applications can improve public participation in resolving transportation issues by helping the public share information and participate in planning and administrative tasks (Nash 2009). Examples of sharing information include the following:

- Wiki Web sites such as StreetsWiki, which allow users to create new pages and edit existing pages, to discuss urban street-related issues

- Blogs, videos on YouTube, online photo albums, and the use of Twitter to share transportation-related information

- Mashup competitions soliciting Web users' creative solutions for transportation. Similar approaches have been demonstrated in the Great Britain Show Us a Better Way mashup competition (`http://www.showusabetterway.com`) and the Washington, D.C., Apps for Democracy mashup competition (`http://www.appsfordemocracy.org/`). Creative uses of Web 2.0 strategies can help foster this kind of citizen participation.

Examples of applications that help government complete a specific task include the following:

- SeeClickFix (`http://www.seeclickfix.com`), a crowd-sourced problem identification application that enables citizens to identify nonemergency issues such as potholes, describe them in detail (e.g., include photos), and place them on a map so other users can add information and to attract attention from the responsible public agency to fix the problem.

- Cyclopath (`http://magic.cyclopath.org`), a crowd-sourced recommendation application developed by the University of Minnesota to allow users to rate the "bikeability" of roads and trails. This geowiki provides an editable map where anyone can share notes about roads and trails, enter tags about special locations, and fix map problems such as missing trails.

- Crowd-sourced planning applications such as the San Jose Wiki Planning Project, which allows users to take a survey, add comments, and post photos, and the Pittsburgh Regional Integrated Transportation Plan, which was actually written through crowd sourcing.

Governments can solicit input and create interest groups through social networking sites such as Facebook and LinkedIn.

All these examples of using Web GIS as a means to solicit public input or VGI and to improve government response to health issues, land-use planning, and transportation management are still pilot projects or research concepts being studied by universities and commercial companies. Nonetheless, the potential for using public input can be applied to many topics of e-government applications. Taking advantage of VGI has become a hot research and application area. USGS, for instance, is actively pursuing VGI as a data source for topographic information. It is using volunteers to update location and attribute information for thousands of point features on the National Map in a collaborative effort with state and federal agencies. It held a workshop in January 2010 to explore ways to incorporate VGI contributions. Other federal agencies, such as the National Oceanic and Atmospheric Administration (NOAA), have also used VGI (*GPS World* 2010). (VGI and public participation GIS [PPGIS] are discussed at greater length in chapter 10.) As Web GIS applications for VGI become more mature and gain wider use, they can contribute significantly to the accuracy of government data and to the responsiveness of government.

## 9.2.3 OPERATIONS AND DECISION SUPPORT

In addition to helping governments become more aware of and responsive to citizens' needs, Web GIS technology has also become an effective tool to keep government agencies running smoothly. Web GIS applications can make current and accurate data available along with powerful GIS tools through a simple user interface that doesn't require a lot of GIS expertise. With the use of Web GIS technology, governments can use the insights provided by placing information in geographical context to engage in better decision making and improved operations. GIS applications provide a solid framework that allows government officials to make informed decisions that have wide-ranging impact on the public.

Web GIS technology furnishes digital tools for abstracting and organizing data, modeling processes, and visualizing information behind an easy-to-use interface that enable leaders to make important decisions without needing extra GIS training. Developments within the last decade in data collection systems, GIS technology, geographic information science, and computing have made the application of GIS methodology to the decision-making process more feasible. Modeling data supplies a method for asking spatial questions that explore the patterns and relationships among various factors. GIS analysis provides greater depth, allowing many layers of data to be considered in aggregate to solve local to global problems. The value of GIS for decision making increases in relation to the scale and complexity of the issue being addressed. Societal challenges such as global warming, resource shortages, and loss of biodiversity, for example, can be better addressed by using GIS technology to sort out competing interests and crucial interdependencies.

Government can improve coordination and communication through the use of a GIS framework that makes complex information easier to visualize and understand. Crucial information can be displayed in a clear and compelling manner with maps, charts, graphs, and models that identify and highlight key concerns. Putting it all online, through the use of Web GIS, increases the flow of information throughout an organization, and to the public.

### MAP-BASED MANAGEMENT DASHBOARD

Government can keep managers and the public better informed by using map-based dashboards. The City of Jacksonville, Florida, has developed a GIS-based online system that monitors government operations in departments throughout the city and speeds up the city response to citizens' requests. Jacksonville, a city with a population of 850,000 and an aerial extent of 841 square miles, faces a complex task in managing city infrastructure and delivery services. The city uses GIS technology in nearly every department and improves operations by using a Web-based system to make the information available to a new class of users: department managers.

JAXGIS, the GIS division in Jacksonville's Information Technology Department, developed the JAXGIS Map Board to enable the City of Jacksonville and Durval County to monitor and manage departmental operations and to provide related information to the community. (The governments of City of Jacksonville and Durval County were consolidated in 1968.) The JAXGIS Map Board is a suite of Web GIS applications used internally by managers and a set of Web sites accessed externally by the public. The JAXGIS Map Board allows managers to keep track of city operations. When a citizen contacts the city to request a service or to make a complaint, whether by phone, e-mail, or online, the incident is assigned an issue number by the Citizen Active Response Effort (CARE) call center. The call is routed to the appropriate department and is also geocoded and added to the JAXGIS map board, which displays dynamic information on the status of citizen

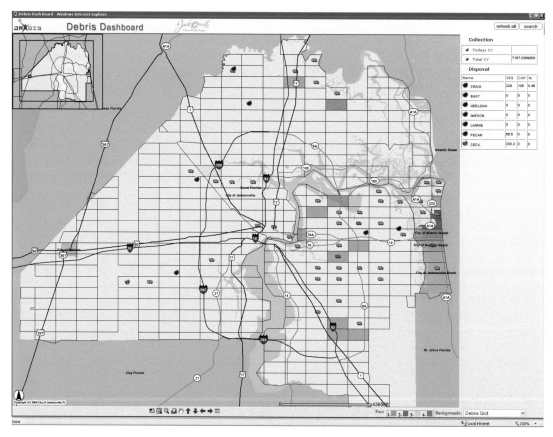

**Figure 9.7** The Debris Dashboard of the JAXGIS Map Board for the City of Jacksonville, Florida, displays dynamic information on the status of citizen calls on waste issues.

Courtesy of City of Jacksonville.

requests color-coded on a map (figure 9.7). For example, when a complaint is directed to the Solid Waste Complaints Status Dashboard, it is grouped by case type (e.g., trash, recycling, yard waste pickup, bulk waste removal), prioritized, and color-coded on a map to show its current status. Case types have estimated completion times, and operations managers can easily view the quantity and distribution of requests, both currently and historically. The JAXGIS Map Board helps managers use Web GIS technology to spot developing problems, reallocate resources, and study city response times. Likewise, residents can use the external JAXGIS Web sites to find information about local properties, view crime incidents by location, learn about businesses and attractions in downtown Jacksonville, and find out about street closures among other things.

Applications like the JAXGIS Map Board improve city operations by enhancing communications among government departments and communications between the city and residents. Web GIS applications improve the amount and quality of information being exchanged and increase the speed of communications between departments and between government agencies. Better communications helps governments avoid a duplication of efforts and helps them to use resources more wisely. It also allows various government agencies to work together to tackle large-scale problems.

## LAND MANAGEMENT

Web GIS plays an integral role in administering federal lands. The National Integrated Land System (NILS) provides a process to collect, maintain, store, and share parcel-based land and survey information that meets the common, shared business needs of land title and land resource management. As a joint project of the BLM, USFS, and states, counties, and private companies, NILS was started in 1998 to provide a model and common tools for managing records on the hundreds of millions of acres of federal lands held in the public trust. NILS provides a series of tools for government to manage land records and cadastral data and to disseminate the information to government employees and the public.

"*The primary purpose of NILS is to automate the BLM cadastral surveying and land records business rules and data in a GIS environment and to produce a national cadastre for the U.S. The system model and the national cadastre are usable by and available to the general surveying community, the land records community, and to the public. The national cadastre provides the BLM, government agencies, and the public with the Public Land Survey System and other survey-based data that is used as a foundation to tie land records to survey-based data in order to improve the positional accuracy of the data in the maps and to make the vertical integration of GIS layers more efficient and less costly.*" (Cone 2008)

NILS comprises four main components: survey management, measurement management, parcel management, and GeoCommunicator. The GeoCommunicator Web site (**www.geocommunicator .gov**) disseminates NILS data and allows agencies and the public to search, access, and dynamically map data pertaining to federal lands, land and mineral use records, and the Public Land Survey System (PLSS). The site contains several interactive mapping applications: the Land and Mineral Use Records map viewer, the Federal Land Stewardship application, and the Land Information System (LIS). The Land and Mineral Use Records map viewer locates mining claims; leasing information on oil, gas, coal, and geothermal sites; mineral use authorizations; and land and mineral titles. Users can link to records from the BLM's LR2000 system or a geographic report that provides detailed information. The Federal Land Stewardship application displays federal surface management agency boundaries, while LIS supplies PLSS data as viewable map layers or downloadable GIS files. The public can get a detailed, accurate look at the nation's lands by using Web GIS technology to access government information.

## EMERGENCY MANAGEMENT AND NATURAL DISASTERS

With an expanding set of GIS services and tools, Web GIS technology promotes collaboration on a scale that can address the issues of a complex world. Coordination of the policies and actions of government at all levels is important to formulating a response to the world's major challenges. Issues such as sustainable development and energy management are formidable in their scope and complexity, requiring a high degree of collaboration among government, businesses, and citizens to reach long-term solutions. Large-scale events, such as hurricanes, tornadoes, and earthquakes, affect multiple jurisdictions, requiring coordination and collaboration on a grand scale. Emergency situations call on government agencies to work together, even when those agencies belong to different levels of government, or to different nations. Because fires, floods, and similar catastrophes know no boundaries, a response typically requires joint activities by agencies of multiple jurisdictions that must coordinate resources, pool information, share analyses, and communicate with the public and the media. Web GIS plays an important role in facilitating collaboration in full-scale emergencies, providing a common operating picture (COP) that is available to all participating agencies.

In a more densely populated world, natural disasters strike larger populations, requiring an even larger and more coordinated response. The Pacific Disaster Center (PDC) in Hawaii, a federal information processing facility that supports emergency managers in the Pacific and Indian oceans regions, exemplifies the power of Web GIS to enhance collaboration in disaster response. The PDC uses Web GIS applications to provide COP to responding agencies for a region that is both large and culturally diverse, covering the Philippines, Thailand, Cambodia, Vietnam, Fiji, Vanuatu, Guam, American Samoa, Hawaii, Alaska, Afghanistan, and the Caribbean. Because disasters in one country can affect surrounding countries, all disaster managers in the region need to be kept informed. Even events that are viewed as local may still require mitigation that involves the entire region. The PDC's use of Web GIS technology goes beyond data collecting. The relief efforts and economic disruption caused by a natural disaster can make these events global concerns, making the use of Web GIS technology even more crucial to the region's disaster managers.

Information flow can be a logistical challenge in any disaster, because responders from different agencies may be using different methods to collect information. Most information is collected according to a specific range of applications. Data collectors tend to structure and think about data exclusively as it relates to a particular field such as transportation or weather, while disaster response requires that all types of information be integrated. The ability to synthesize all types of data is what makes GIS such an effective tool. The PDC applies expertise from many different disciplines to supply geospatial and imagery resources to the region's disaster managers, and it provides customer support through the use of Web GIS to help managers develop the kinds of applications needed to respond to large-scale emergencies.

Other examples of how Web GIS technology can be used to improve government decision making are cited elsewhere in this book in chapters 5, 6, and 7. This chapter summarized three main types of Web GIS applications: (1) public information services, which are the most visible to the public; (2) two-way interactive communications, a hot research area with great potential for promoting public involvement in government affairs; and (3) operations and decision support, which is mainly used internally by governments.

## 9.3 CHALLENGES AND PROSPECTS

Governments' needs for GIS have been one of the main drivers of the research and development of GIS. To be successful, the design of e-government applications needs to account for a number of factors, including interoperability and scalability. Research on Web GIS applications in government are closely associated with research in neogeography, VGI, PPGIS, geocollaboration, information sharing, cloud computing, and the integration of emerging Web 2.0 technologies within e-government applications. Outcomes of this research promise to make Web GIS technology an increasingly stronger foundation for the future of e-government.

### 9.3.1. DESIGN CONSIDERATIONS

While many of the design considerations for Web GIS have been discussed in chapter 2 (designing the user experience) and chapter 3 (optimizing Web services), the design of e-government applications calls for special attention to the following considerations:

- **Capacity planning:** Government Web sites serve the needs of citizens and government employees, who can number from the hundreds to the hundreds of thousands, so scalability is an important design consideration. As government Web sites incorporate functionality to carry out transactions, these applications need to perform reliably, while safeguarding citizens' personal and financial information. If transactions fail or information is misappropriated, citizens lose confidence, making these online applications ineffective. Current and future requirements for bandwidth, databases, Web servers, and GIS servers need to be considered when designing Web GIS applications for e-government.

- **Security:** Web GIS, as the means of delivering government services via e-government, suffers from the same criticisms that are generally directed at the Web. Applications that involve government interactions with the public and with commercial organizations often involve citizen or business private information, while applications that involve government staff often have information that should be kept internal. Concerns over data security and individual privacy extend to Web GIS applications that are used to carry out government transactions. Effective security, including the controls introduced in sections 3.5.5, 7.4.2, and 8.3.1, needs to be enforced to protect citizen privacy and internal government information.

- **Interoperability:** Government agencies use a variety of software tools. Adherence to industry standards, especially standards at the Web services level, is crucial for a smooth flow of geospatial information across organizations and across platforms.

- **Extensibility:** Rapidly changing Web technology mandates that Web GIS applications should be designed with extensibility in mind and leave room for future growth such as adding new functionalities, using new data formats, adding or replacing hardware and software, and adding new users.

## 9.3.2 THE FUTURE OF WEB GIS IN E-GOVERNMENT

Web GIS is making GIS technology widely available to government agencies and citizens. Web GIS has substantially affected the way government agencies operate, from the ways they interact to the ways they serve the public. Despite the progress, however, the potential of GIS as a valuable support for e-government is not being fully realized. Further progress for Web GIS applications in e-government will depend on a number of factors.

Constraints on access to data remain one of the greatest challenges to extending the use of Web GIS in government. Without appropriate data, the most sophisticated applications are of no practical use. Government agencies are the largest producers of data as well as the largest consumers of geospatial data, but the producers and the consumers of a piece of data are not always one and the same. One agency often needs access to another agency's data. Obtaining data by duplicating datasets is often not the most effective in terms of the efforts involved and the time delays. Government can leverage its huge investments in data through the development of geoservices, GIS functionality delivered as Web services, that can connect government to government and government to citizens. The result will be a geospatial framework for Gov 2.0. Jack Dangermond, president of ESRI, characterized the effect of geoservices by saying, "Exposing these services will bring about as much change as GIS itself has brought. Existing services could be combined into new services. This would remove impediments between organizations currently using geospatial data and allow them to work in a loosely coupled environment that favors collaborating and encourages synergies" (Pratt 2009). The use of Web services underpins work on the next-generation NSDI, or NSDI 2.0 (see chapter 7), and the future of e-government.

Cloud computing technology is a major feature of President Obama's initiatives to modernize government IT. A significant percentage of government investment in IT goes to maintaining infrastructure each year, according to chief information officer Vivek Kundra. By using commercially available technology, cloud computing helps government to focus on delivering value-added applications rather than spending time on managing infrastructure (Kundra 2009). Properly employing cloud computing, including cloud-based GIS, in the development of e-government brings the prospect of cost savings, greater efficiency, and innovative applications.

Web GIS applications in government are part of a larger evolution. Instead of merely extending the reach of GIS on the desktop, Web GIS is a platform for fostering collaboration and innovation across the Web—connecting people and businesses with government and each other. Tim O'Reilly, founder and CEO of O'Reilly Media, has dubbed this transformation Gov 2.0 (2009), along the lines of Web 2.0. While there have been pilot projects that scratch the surface, research on how to seamlessly integrate Web 2.0 technologies such as social networking with Web GIS, allowing interactions with citizens and staff to solve real-world problems, and protecting individual privacy at the same time, are topics that deserve attention.

The application of Web GIS in government is closely associated with a number of research areas such as how to validate the quality of VGI and employ it for authoritative uses, how PPGIS can best empower and encourage citizens to expand public involvement, how geocollaboration can efficiently support remote and synchronous collaboration, how the mostly visual use of virtual global and 3D visualization can be extended for problem solving and decision support, and what implications the popularity of online games such as virtual worlds have for e-government. Principles and theories that come out of this research can guide enhanced applications of Web GIS in e-government.

Web GIS applications have clearly demonstrated the value of geography and GIS in improving government services and supporting decision making. The insights provided by viewing data in a spatial context are directly accessible to policy makers at all levels of government and to citizens through a simple Web interface. While government has often been maligned for hiding government processes, Web GIS applications enhance the transparency of government operations and make it easier to measure the outcomes of government programs. Government is increasingly employing the tools of social networking and Web GIS to make government processes more transparent and more responsive. Government becomes more participatory because communication is bidirectional, and decision making gains new immediacy. As the underlying technologies of the Web and GIS continue to evolve, and as geospatial data and Web services are increasingly being shared and becoming more accessible, Web GIS will extend its reach and offer stronger support for the future of e-government.

## Study questions

1. What is e-government or e-gov?

2. How did GIS help government agencies before Web GIS? What benefits does Web GIS bring to government applications?

3. Summarize several types of applications of Web GIS in e-government.

4. Describe a Web GIS application that is used for public information services. List the objectives, functions, and technologies used in this application.

5. Think of a scenario where government can use Web GIS to collect public opinions. What benefits does Web GIS produce in this scenario?

6. Emergency response often needs to integrate multiple Web resources and have the means to collaborate among multiple agencies and communicate with crews out in the field. What capabilities can Web GIS provide? Without Web GIS, how will emergency response be affected?

7. Describe several design considerations that should be accounted for in Web GIS for government projects. Why? How?

8. Has your city or county used Web GIS to deliver services or support operations? If so, what functions are delivered or supported? If not, what specific functions can Web GIS provide for your local government?

## References

*ArcNews*. 2006. Integrated cadastral system for U.S. Public Land System: BLM's GeoCommunicator Web site accesses and distributes spatial data. *ArcNews Online* (Spring). http://www.esri.com/news/arcnews/spring06articles/blms-geocommunicator.html.

———. 2009. Maryland's programs improve life and environment: Governor Martin O'Malley leads with GIS. *ArcNews* (Summer).

CEQ NEPAnet. 1970. The National Environmental Policy Act of 1969. http://ceq.hss.doe.gov/nepa/regs/nepa/nepaeqia.htm (accessed January 29, 2010).

City of Baltimore. 2004. CitiStat. http://www.innovations.harvard.edu/awards.html?id=3638 (accessed March 15, 2010).

Clinton, William J., and Al Gore. 1997. *The Blair House papers*. Washington, D.C.: National Performance Review.

Cone, Leslie M. 2008. Explaining the National Integrated Land System (NILS). Tenth International Conference for Spatial Data Infrastructure, February 25–29, St. Augustine, Trinidad. http://www.gsdi.org/gsdiconf/gsdi10/papers/TS37.3paper.pdf (accessed January 29, 2010).

Eggers, William D. 2005. Government 2.0: Using technology to improve education, cut red tape, reduce gridlock, and enhance democracy. Lanham, Md.: Rowman & Littlefield.

ESRI. 2007. GIS integration with public safety applications. An ESRI white paper. Redlands, Calif.: ESRI.

Gartner. 2002. Trends in U.S. state and local governments. *Market trends* (March 19): 4.

Goodchild, Michael. 2007. Citizens as sensors: The world of volunteered geography. http://www.ncgia.ucsb.edu/projects/vgi/docs/position/Goodchild_VGI2007.pdf (accessed January 29, 2010).

*GPS World*. 2010. Volunteered geographic information for the National Map. http://www.gpsworld.com/gis/news/volunteered-geographic-information-the-national-map-9429 (accessed January 29, 2010).

Kundra, Vivek. 2009. Streaming at 1:00: In the Cloud. The White House Blog. http://www.whitehouse.gov/blog/streaming-at-100-in-the-cloud (accessed March 15, 2010).

Nash, Andrew. 2009. Web 2.0 applications for improving public participation in transport planning. http://www.andynash.com/nash-publications/2009-Nash-Web2forPT-14nov09.pdf (accessed December 9, 2009).

Obama, Barack. 2009. Transparency and open government: Memorandum for the heads of executive departments and agencies. http://www.whitehouse.gov/the_press_office/Transparency_and_Open_Government (accessed March 12, 2010).

O'Reilly, Tim. 2009. Gov 2.0: It's all about the platform. *TechCrunch*, September 4. http://www.techcrunch.com/2009/09/04/gov-20-its-all-about-the-platform/.

Osborne, David, and Ted Gaebler. 1992. *Reinventing government: How the entrepreneurial spirit is transforming the public sector*. Reading, Mass.: Addison-Wesley.

Page, Doug. 2006. GIS: Bringing disaster management down to earth. *Homeland Protection Professional*, June 15. http://www.homeland1.com/homeland-security-columnists/doug-page/articles/350133-gis-bringing-disaster-management-down-to-earth/ (accessed August 29, 2009).

Pratt, Monica. 2000. Conference focuses on the global network. *ArcUser Online*. http://www.esri.com/news/arcuser/1000/uc2000.html (accessed March 3, 2010).

———. 2001. Center integrates response for half the globe. *ArcUser Online*, July–September. http://www.esri.com/news/arcuser/0701/pdc.html (accessed January 29, 2010).

———. 2006. The GeoWeb: A vision for supporting collaboration. *ArcUser*, Jan–March. http://www.esri.com/news/arcuser/2006/geoweb.html.

———. 2009. The next step: The importance of building geospatial infrastructures. *ArcUser* (Spring): 16.

Thomas, Christopher, and Milton Ospina, eds. 2004. *Measuring up: The business case for GIS*. Redlands, Calif.: ESRI Press.

U.S. 107th Congress. 2002. E-Government Act of 2002. Public law 107–347. Dec. 17, 2002. `http://frwebgate.access.gpo` `.gov/cgi-bin/getdoc.cgi?dbname=107_cong_publiclaws&docid=publ347.107.pdf` (accessed March 15, 2010).

U.S. Census Bureau. 2004. Census Bureau marks 10th anniversary of `www.census.gov`; serves 1.5 million pages to Web surfers per day. `http://www.census.gov/webdecade/`.

Walsh, Trudy. 2009. Is a national GIS on the map? *Government Computer News*, July 13. `http://www.gcn.com/` `Articles/2009/07/13/National-GIS-Federal-agencies-ESRI.aspx?Page=4` (accessed August 26, 2009).

Walters, Debra. 1999. Sacramento serves public timely information over the Web. *ArcUser* (April–June): 37–8.

Zulick, Carl. 2004. E-government for planning and environmental issues. In *Measuring up: The business case for GIS,* eds. Christopher Thomas and Milton Ospina. Redlands, Calif.: ESRI Press.

# HOT TOPICS AND NEW FRONTIERS

It sounds like a cliché to say that the Web has changed the way we live and work. Electricity, land-line telephones, and television changed our lives. But the creation of the Internet and the Web is another paradigm shift that is changing our way of life at a faster pace than ever before. The development of the Web has been phenomenal. There is no other technology that has been so rapidly adopted by such a large number of people. Web GIS, the combination of the Web and GIS, is advancing rapidly since its inception in 1993. It is creating many new research areas and renewing existing avenues of research, attracting attention from both the academic and business communities.

The first section of this chapter introduces Web GIS-related research hot spots and frontiers, including volunteered geographic information (VGI), public participation GIS (PPGIS), collaborative GIS, geotagging, geoparsing, geotargeting, online virtual reality, the sensor Web and sensor network, the Semantic Web, and cloud-based GIS. The second section introduces the characteristics that will define the next generation of the Internet and the Web, and predicts the future trends of Web GIS.

## 10.1 AREAS OF RESEARCH

This section introduces several new research areas created by Web GIS and looks at existing research areas in which Web GIS can provide enhanced solutions.

### 10.1.1 VGI

User-generated content (UGC) is a significant phenomenon in the Web 2.0 era, and volunteered geographic information (VGI) is UGC of a geospatial nature. The massive amount of content that has been contributed by millions of Web users is responsible for the mass appeal and success of Wikipedia, YouTube, MySpace, Facebook, Second Life, and many other popular Web sites. In 2004, ABC News even listed bloggers as People of the Year. In 2006, Business 2.0 ranked Web users as the first in its list of "50 people who matter now," and *Time* magazine chose Web users as its Person of the Year. **VGI is digital spatial data that is created voluntarily by citizens rather than by formal data producers** (Goodchild 2007a). Web sites such as WikiMapia, OpenStreetMap, and SeeClickFix use Web-based "crowd-sourcing" technologies, which outsource tasks to the crowd—that is, mostly to the large group of Web users—to collect data, identify problems, and invite solutions to problems through the use of VGI. The use cases, essential geospatial questions asked, and information collected by many popular Web sites that facilitate VGI are categorized in table 10.1.

| EXAMPLE WEB SITES AND ESSENTIAL GEOSPATIAL QUESTIONS ASKED | USE CASES | GEOGRAPHIC INFORMATION CONTRIBUTED |
|---|---|---|
| WikiMapia<br>*What places do you know?* | Users draw a rectangle and describe the place with a few sentences. | Constructing a comprehensive global gazetteer database |
| Picasa, Panoramio, and Flickr online albums (geotagging)<br>*What photos can you share about places you have been?* | Users upload geotagged photos or upload and geotag the photos by zooming to a location on a map. | Recording and reporting the past and present conditions of places or events with the use of photos |
| OpenStreetMap<br>*What GPS data do you have for the roads you bike, walk, or drive?* | Users upload the track logs of their personal GPS receivers. | Constructing street and highway data layers for many areas |
| SeeClickFix<br>*Where and what problems do you see that need to be fixed?* | Users report problems such as potholes and graffiti by drawing on and annotating maps. | Identifying problems for 2-1-1, 3-1-1, 4-1-1, and 9-1-1 services |
| Twitter (GeoTweeting)<br>*What's going on there?* | Users report personal activities, other events, or incidents at their locations. | Monitoring and reporting activities, incidents, or events in real time |

**Table 10.1**   Web sites that facilitate and harvest VGI

Many of these functions can be implemented by using Web editing capabilities such as those supported by ArcGIS Server feature editing services (see chapter 3). Some applications are introduced in chapter 7 (reporting graffiti) and in this chapter (photo geotagging). The GeoTweeting

**Figure 10.1**  Using an ArcGIS Explorer Desktop Twitter add-in, found at `http://www.arcgis.com`, you can submit GeoTweets by clicking a location on the map. You can also map your friends' GeoTweets.

Courtesy of U.S. Geological Survey and Twitter.

feature launched by Twitter is another example that facilitates VGI (figure 10.1). Twitter is a social networking and micro-blogging service that gives users the ability to send or receive short (140-character) text messages known as "tweets." This can be accomplished from a wide variety of Twitter client applications available on a range of platforms, including mobile phones. With the addition of geolocation support, users now have the ability to send information (e.g., latitude and longitude) with each message. This is known as "geotagging your tweet," or "GeoTweeting" for short. For example:

"*Three-car collision on I-10 eastbound near the Alabama exit. Traffic stopped. You'd better avoid this area for the next hour.*

*<34.0666, -117.20872>*"

Some Twitter client applications support GeoTweeting by automatically obtaining your location from the GPS on your mobile phone, allowing you to specify a location on a map or to type in your latitude and longitude. For example, the ArcGIS Explorer Twitter add-in allows you to submit GeoTweets as well as visualize your friends' GeoTweets (LaFramboise 2009).

VGI provides a unique value and perspective to global observation, NSDI development, and public participation GIS in the following ways:

- **People acting as sensors:** In the Web 2.0 era, a significant percentage of the world's 6 billion people can use their senses as well as their mobile phones, cameras, camcorders, and other tools to collect geospatial information and share it. VGI provides a human dimension that aerospace remote-sensing monitoring systems don't provide. People can see details that satellites don't see, and they have more constant access to places that satellites reach only periodically. VGI has the potential to be a significant source of information for a variety of Web GIS applications, including real-time disaster monitoring and early warning.

- **Enormous opportunities for SDI development:** Traditional data collecting requires appropriately trained personnel and mechanisms for communication and dissemination. This restriction has resulted in a notable decline in the production of geographic information by many governments (Goodchild, Fu, and Rich 2007; Goodchild 2007b). Yet it is irrelevant to VGI since public participation and UGC supply an increasingly exhaustive amount of data for free.

- **Shedding light on PPGIS:** Why do citizens voluntarily contribute their observations, data, and knowledge on their own time for free? Understanding the reasons behind VGI will aid the development of PPGIS, which aims to motivate citizens to participate in decision making.

While VGI is of significant value, it has also prompted concerns over its quality and integrity —that is, its accuracy. Truth in content is asserted by the person who creates it without regard to citation, reference, or other authority. The mechanisms for ensuring the validity of VGI and of protecting personal privacy are important areas for further research.

## 10.1.2 PPGIS

**The goal of PPGIS is to involve the general public in producing needed data and participating in decision making through the use of GIS.** The term "public participation GIS" was coined in 1996 at the National Center for Geographic Information and Analysis (NCGIA) conference (Sieber 2006), but the concept is believed to have originated much earlier. Many government activities, such as regional and local planning, have a direct bearing on citizens. Citizens have a right to know what's going to happen in their communities and the right to voice their opinions. More importantly, citizens have immense knowledge about their environs, so they can help governments make better decisions. PPGIS aims to facilitate public participation through geographic technology, including digital maps, satellite imagery, and map sketches. This goal of encouraging public participation in decision making affecting the community makes PPGIS an important element that can contribute to the advancement of government 2.0.

PPGIS faces both social and technical barriers. Some citizens might not find it convenient to attend public hearings at a certain place and time. Other citizens might not feel comfortable speaking up about the issues in a public forum. Some people might find it difficult to access and use the relevant data, maps, and tools. Web GIS can potentially remove many of these barriers, but its potential has not been fully realized, in large part because of challenges in the design and usability of Web GIS applications. Thus, most PPGIS projects have been experiments or pilot studies instead of practical applications (Batty 2007).

Web 2.0 provides a new suite of principles and technologies that can facilitate public participation by encouraging people to use the Web through a good user experience that can engage the public (Nash 2009). Web maps, cloud computing, and geobrowsers provide easy access to free data, maps, and tools; social networking Web sites connect the public, the community, and government; and Web-based crowd-sourcing technologies leverage public collaboration to tackle tasks traditionally performed by commercial vendors. PPGIS can involve the public by using the following techniques:

- **Web-map-based approach:** Many of the Web sites that harvest VGI can use maps to reach the public. Citizens can sit comfortably at home, browse online maps to see planned zoning or planned projects around their neighborhood, click or sketch on the map and describe the conditions or problems they see, and then submit their opinions or recommendations online without having to go to city hall and spend hours to attend a public hearing (figure 10.2).

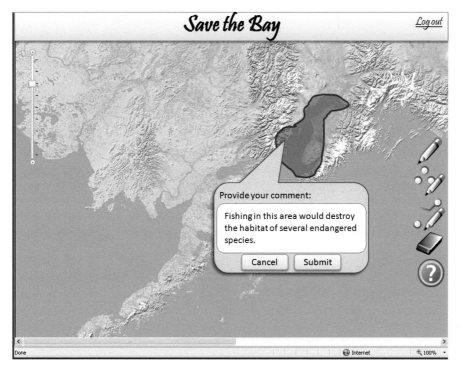

**Figure 10.2** A mockup of a draw-and-comment Web application invites the public to make recommendations on a save-the-bay project.

Courtesy of U.S. National Park Service.

- **Blog and video/photo-sharing Web sites:** Governments can set up Web sites that use easy-to-understand blogs and other media to encourage citizens to discuss the issues and leave comments.

- **Social networking approach:** Governments can create groups that include local communities by using Facebook, Twitter, or other social networking applications to share information and post questions such as "What are your ideas to improve the environment in our city?"

Such resources have to be well structured and well designed to provide an engaging user experience. Social considerations must be taken into account as well as technical considerations. Governments can stimulate citizen awareness by showing how policies will affect their communities. They can provide incentives for participation by recognizing citizens who provide input or giving out prizes for suggestions.

### 10.1.3 GEOCOLLABORATION

**Geocollaboration, or collaborative GIS, means a group of people using GIS technology to create a collaborative working environment to complete the same task.** Geocollaboration has both spatial and temporal characteristics (MacEachren 2000), namely the following:

- Spatially, the collaborators can be either colocated, for example, standing around a map and talking about a new zoning plan, or remotely located, for example, talking on the telephone with each person having a map in front of them.

- Temporally, the collaboration can happen synchronously (in real time) or asynchronously (at different times).

Geocollaboration is a new field of research that investigates how geospatial technology can be used to support person-to-person collaboration and decision making. The demand for collaborative GIS is increasing, leaning more toward remote and synchronous collaboration. A lot of planning and decision making, especially in emergency response, is carried out through the collaboration of multiple groups or multiple agencies that are not all sitting in the same room, and may even be spread out among agencies across the globe. Given this geographic disparity, a GIS that supports remote collaboration typically has five system characteristics (MacEachren, Brewer, and Steiner 2001; Brewer et al. 2000):

1. Facilitating dialog: ability to talk or chat while viewing and interacting with the GIS

2. Accounting for group behavior: ability to know what others are doing

3. Drawing the group's attention: ability to indicate objects, places, and regions, and to alert others to the conditions

4. Allowing private work: ability to work out ideas individually before sharing them with others

5. Allowing saved and shared sessions: ability to save and initiate new analysis from any point, which is especially important for asynchronous collaboration

Advances in Web GIS have made dramatic improvements in remote interaction involving geography (figure 10.3). In essence, geocollaboration is about sharing and exchanging data (maps, drawings, annotations, messages, alerts, etc.), often in real time, and it can be among all participants or just between certain participants. In chapter 5, in a collaboration scenario for emergency

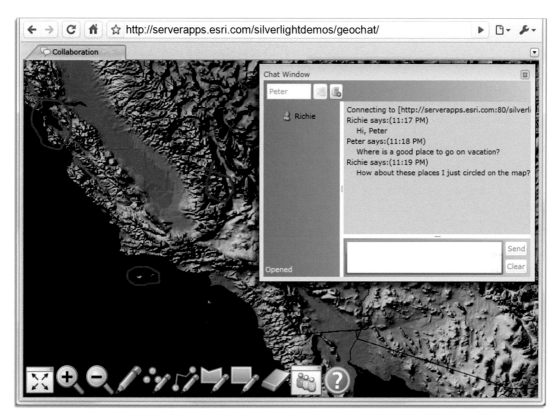

**Figure 10.3** A Web-map-based whiteboard and online chatting application allows people at different locations to instantly conduct a map-based conversation.

Courtesy of U.S. Geological Survey.

management, a first responder reports a chemical spill using a mobile GIS client. The information is passed to the GIS server at the emergency operations center (EOC). The EOC runs a plume geoprocessing model to calculate the spatial area that will be impacted and to identify the neighborhoods that need to be evacuated, the roads that need to be blocked, and the shelters that need to be used. The resulting map is passed to the police department so it can swiftly and securely close certain roads, to the health department so it can plan ambulance services, and to other key emergency response agencies as needed. Geocollaboration enhances situational awareness and provides participants with a common operating picture (COP) that lets everyone "see" the same information simultaneously. The group Short Messaging Service (SMS) function described in chapter 5 is also an effective way for the EOC to initiate dialog and broadcast commands to all field crews in a certain area.

Geocollaboration is at the confluence of the Web, GIS technology, and computer-supported cooperative work (CSCW) (Schafer et al. 2005). The most widely used and appreciated CSCW software, also known as collaborative software or groupware, includes instant messaging (i.e., real-time text chat), voice and video conferencing, Web conferencing (i.e., sharing a desktop screen), whiteboards, blogs, SMS, Twitter, Web calendars, and e-mail. These technologies make collaborators more available to one another and help to integrate synchronous and asynchronous communication models and joint work. By integrating Web GIS with these groupware technologies or implementing similar groupware in the map context, data such as geometries, labels, text messages, maps, modeling results, audio, and video can be swiftly and easily exchanged. For example, using a map-based whiteboard and online chatting (Carmichael 2009), two friends can discuss good vacation spots. One can recommend places to go by drawing them on the map. The other person, who sits somewhere else using the same Web map application, can immediately see the places his friend recommended and draw his own recommendations as well.

The possibilities for geocollaboration are promising, where virtual teams can work together on problems and solutions without having to meet face to face. Geocollaboration can build virtual EOCs that allow groups in different locales to coordinate complex real-time emergency response. Geocollaboration can save time and money by allowing team members to come together in a virtual setting to make fast and accurate assessments in real time over the Internet. The use of a common virtual workspace can reduce staff travel time and corporate expenses by providing communication channels and the means of interaction in support of real-time awareness. Geospatial intelligence can be effectively integrated into various decision-making processes to establish a common operating picture that fulfills the purpose of geocollaboration.

### 10.1.4 GEOTAGGING

**Geotagging is the process of adding location information to various digital media such as photos, videos, Web sites, and RSS feeds.** Geotagging defines location, which typically consists of latitude and longitude coordinates, although it can also include altitude and bearing. Geotagging photos is the most prevalent type of geotagging.

The location of photos can be geotagged automatically or manually:

- **Automatic geotagging:** Many digital cameras come with built-in GPS receivers, which automatically geotag photos. Cameras by traditional manufacturers such as Nikon, Canon, and Ricoh include GPS receivers as do some cell phone cameras, including iPhone, Blackberry, Samsung Memoir, HTC EVO 4G, and a variety of other models. Some cameras even have a built-in compass so that the latitude and longitude of a photo, along with the compass bearing, are automatically written into the image file when the photo is taken. The latitude and longitude coordinates are typically in WGS84 (World Geodetic System 1984). For photos in JPEG format, the geotag information is typically embedded in the metadata, in EXIF (exchangeable image file) or XMP (Extensible Metadata Platform format). This data is not visible in the picture itself but is readable by many photo processing programs and most digital cameras.

- **Manual geotagging:** Photos taken by a camera without a GPS receiver can be geotagged manually. There are many tools, such as ArcGIS Explorer Image Geotagger (discoverable on ArcGIS Online), Microsoft Expression Media, and tools provided by Picasa, Panoramio, and Flickr, that allow users to manually geotag photos by zooming to the location on a Web map where the photo was taken (figure 10.4).

**Figure 10.4** With an image geotagger add-in tool available via ArcGIS Online, ArcGIS Explorer Desktop allows users to geotag photos, left, as well as display geotagged photos, right.

Courtesy of U.S. Geological Survey.

Geotagging adds value to GIS in the following ways:

- **Spatially organizing photos and other types of data:** Photos are routinely collected for various purposes such as personal hobbies, public works documents, and historical records. If not geotagged, the exact location that these photos represent and the information that is recorded on them (i.e., buildings, streets, pipes, rivers, hills, and so on) will either remain unknown, which will make it difficult to discover the photos based on location, or require that the information be recorded separately, which will introduce extra work in terms of documentation and management. Geotagging permits easy and accurate positioning of photos in a spatial context, as well as allowing a quick search for photos taken near a given location or facility.

- **Enriching geographic information:** Geographic information largely consists of geometry and text attributes. Adding geotagged photos and other media makes GIS more intuitive, vivid, and engaging.

- **Providing valuable information for data mining:** Web users can view geotagged photos in many Web map applications. With billions of photos geotagged and uploaded to the Web, users can derive much valuable information from visualizing these photos spatially and analyzing their spatial patterns.

### 10.1.5 GEOPARSING

**Geoparsing is the process of assigning geographic location information to textual words and phrases in a document.** Geoparsing differs from geotagging in that in geoparsing—which is mainly for text documents such as Web pages—each paragraph, or even each sentence, of a Web page could be assigned a separate location, whereas in geotagging—which is often for photos and videos even when it is used for Web pages—the whole Web page would be assigned one location. A **geoparser** is a piece of software or a (Web) service that does geoparsing. Such products include Geographic Search and Referencing Products (GSRP) by MetaCarta and GeoLocator by Digital Reasoning. The geoparsing results can be inserted into the original document, used to produce a new document, or formatted for output to a geospatial application. For example, chapter 4 illustrates a Web service by GeoNames that can convert a news RSS into a GeoRSS. The process behind this conversion is geoparsing.

Geoparsing involves geospatial entity extraction that identifies place-names and disambiguation that georeferences the place-names (Caldwell 2009) as follows:

- Geospatial entity extraction processes the natural language to identify place-names in the text. For example, in the following sentence, there are multiple place-names:

> *"Mr. Washington stood at the corner of Alabama Street and State Street in Redlands, eating a Chicago-style hot dog."*

The geoparsing software must analyze the context and grammar of a sentence and identify the words that form place-names. The software should recognize that Washington

- in *Washington state* refers to the name of a place;
- in *George Washington* refers to the name of a person;
- in *Washington Post* refers to the name of a newspaper;
- in *Washington Mutual* refers to the name of a bank.

Ideally, geoparsing software should also understand relative positions such as *"50 miles east of Los Angeles"* and *"northwest states of the United States"* and assign proper spatial extents to them.

- Disambiguation finds the correct location for the identified place-names. This step searches a gazetteer—that is, a directory of places and their geographic extents—to convert the place-name into its coordinates. Disambiguation can be challenging, because place-names are often duplicated in locations across the country. In this example, the duplicated place-names for Washington pose an issue. The Geographic Names Information System (GNIS), the official repository of U.S. geographic names data, returns more than 2,000 names across the country that match Washington. The software would have to associate the correct Washington based on additional context in the text document.

Geospatial entity extraction and disambiguation are vulnerable to the vagaries of language. Current research on the Semantic Web, and indeed the future realization of the Semantic Web, aims to make geoparsing easier and more accurate.

**Geoparsing can turn text documents into a geospatial database. It can organize mass volumes of text documents spatially, permit their contents to be plotted on a map, and allow a quick search for information buried in the text documents.** This is important for missions such as antiterrorism and military intelligence that need to search for references to a specific geographic location in large volumes of Web pages, e-mails, news feeds, reports, presentations, and many other document types. For example, to identify a possible terrorist attack on a certain city, every piece of information about that city may provide clues that can reveal a threat. If it weren't for geoparsing, analysts would have to spend many hours performing text searches and manually reading these documents to look for direct and indirect references to identify the location and extent of the threat. But it is a process that should be done as quickly as possible to prevent the attack and save lives. Geoparsed documents that combine text search and geographic search capabilities allow analysts to see and find the information easier and faster (Ridley, Gross, and Frank 2005).

### 10.1.6 GEOTARGETING

**Geotargeting is a method for determining the physical location of a visitor to a Web site and delivering tailored content to that visitor based on his or her location.** For example, an advertising Web site can deliver ads about a warm coat to a person who is in Alaska and ads about a swimsuit to another Web site visitor who is in Southern California (figure 10.5).

There are multiple ways to obtain a Web user's location (table 10.2):

- **By the user's registration information:** Some e-mail and social networking Web sites ask users to provide their location information and can later target the user with contents that pertain to the user's location. However, it's worthwhile to note that 20 to 30 percent of Web users typically provide false information to protect their privacy (Carmagnola and Cena 2009; Gellman 2002).

- **By the user's IP address:** When a user visits a Web site, the public IP address of his or her computer is sent to the Web server and can be used by the Web server to determine his or her location. A public IP address of a computer over the Internet is similar to the street address of a

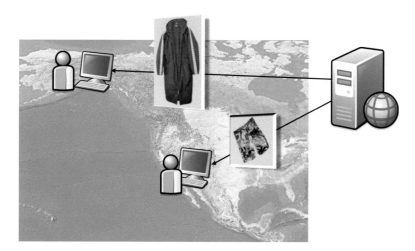

**Figure 10.5** A Web server can deliver appropriate contents, from winter coats to swimsuits, to Web visitors from different places through the use of geotargeting.

Courtesy of U.S. National Park Service.

house. Similar to apartments in the same building that share the same street address, computers with the same ISP (Internet service provider) often share one or several public IP addresses. The spatial resolution of locations obtained from this method typically can only get to the city level, but this is enough for many geotargeting applications. One notable advantage of this approach is that a user's location is automatically determined in real time rather than having to ask the user for the information.

- **From the GPS receiver of the user's mobile phone:** GPS receivers provide accurate location information, which can be used to deliver highly accurate geotargeting services.

| METHODS TO DETERMINE LOCATION | ADVANTAGES | DISADVANTAGES |
|---|---|---|
| By the user's registration information | Users provide the information during registration | • Have to ask users to register<br>• Can be incorrect as some users provide false location information |
| By the user's IP address | Can obtain users' location automatically in real time, without having to ask users for their location | • The spatial resolution is not very high (typically city level, but this is often sufficient for many applications)<br>• Can be wrong in a small number of cases, such as when users use proxy servers or when hackers use IP spoofing to provide a false IP address |
| By the GPS receiver of the user's mobile phone | High accuracy, to the meter | • Only available to devices with GPS<br>• The line of sight required for a GPS receiver to work can be blocked by surrounding buildings or terrain |

**Table 10.2**   Obtaining a Web user's location for geotargeting

Geotargeting can be useful in a variety of applications:

- **Precise advertising based on location:** This is the most common use of geotargeting. As illustrated in figure 10.5, the user in Alaska is more likely to buy a warm coat than a swimsuit, and vice versa for the user who comes from the beaches of Southern California.

- **Fraud prevention in e-commerce payments:** A Web site can identify suspicious payment transactions for online purchases by comparing the location of the user represented by his or her IP address, the home address of the owner of the credit card used to pay for the purchase, and the address of the purchase delivery.

- **Compliance with legal or license regulations:** Legal and license regulations may require that online products or services be delivered only to certain countries or regions, and not beyond these places. For example, during the 2008 Beijing Summer Olympics, the video broadcasting rights in the United States were acquired exclusively by NBC. It would have been illegal for other companies to relay the videos to the United States. Sina.com, a Chinese Web site, was able to comply with the legal regulation by using geotargeting technology (figure 10.6). While it provided real-time Olympics videos to Web site visitors in mainland China, it declined Web users from the United States. In this case, Sina.com determined the users' locations by their IP addresses and decided whether it could provide the video relay services to Web site visitors.

**Figure 10.6** Companies other than NBC were not allowed to broadcast real-time 2008 Summer Olympics videos to audiences in the United States. Chinese Web sites such as Sina.com complied with the regulation by using geotargeting to decline services to Web visitors from the United States.

Courtesy of SINA.

## 10.1.7 ONLINE VR

**Virtual reality (VR) is a technology that allows a user to interact with a computer-simulated environment, whether that environment is a simulation of the real world or an imaginary world.** VR is essentially related to the concept of "telepresence," the sense of being immersed in an environment generated by natural or mediated means. Telepresence is created with two technological dimensions: vividness and interactivity (Steuer 1993).

- **Vividness:** the ability of a technology to produce a sensor-rich mediated environment. The two main factors contributing to vividness are sensory breadth, which refers to the number of sensory dimensions presented, and sensory depth, which refers to the resolution within each of these perceptual channels.

- **Interactivity:** the extent to which users can participate in modifying the form and content of a mediated environment in real time. The three main factors that contribute to interactivity are speed, meaning the rate at which input can be assimilated into the mediated environment; range, meaning the number of possibilities for action at any given time; and mapping, meaning the ability of a system to map its controls to changes in the mediated environment in a natural and predictable manner.

When VR technology emerged in the 1980s, it required high-end computers with superior graphics cards, but most computers today are capable of running VR software. VR file standards have also progressed from VRML (Virtual Reality Modeling Language) to X3D (extensible 3D). Today, many VR applications can run smoothly in Web browsers and other Web clients, resulting in **Online VR**, which is accessible to a much broader audience. VR can have the following forms:

- **3D objects or surfaces that the user can rotate to see from various angles:** This is the most common form. Examples include virtual globes such as ArcGIS Explorer and Google Earth (see chapter 2), which are a 3D model or representation of the Earth or another world with which users can interact by changing the viewing angle and position.

- **3D structures that users can enter and navigate in and around:** Examples include 3D buildings, tunnels, and rooms that users can enter and tour (figure 10.7).

- **Online virtual worlds:** A virtual world is a computer-based simulated environment intended for users to inhabit and interact via avatars—that is, their graphical representations. Examples of online virtual worlds include Second Life, Active Worlds, World of Warcraft, and Kaneva. Virtual worlds usually have a geographic setup such as an island or a city. Web users can choose places to build and decorate their own 3D VR homes, shop online in virtual malls and chat with store clerks, play games, make friends, and chat with other avatars.

- **High-end specialized VR:** Such VRs can include additional sensory information, such as sound and haptic systems with tactile feeling. Players wear special goggles and gloves that make the illusion of reality more complete. Popular games such as Nintendo's Wii sports and Wii fitness games allow users to play golf on a chosen course or to run in chosen scenic locales using Wii remote controls. Players must account for the ground cover, wind direction and speed, and ground slope in their VR activities. Other examples include military training and combat drill systems. Most high-end VRs are custom made and are not necessarily designed for online use.

**Figure 10.7**   3D VR structures allow users to step in the door and tour inside.

Courtesy of B-Design 3D Ltd. and Pictometry International Corp.

VR is closely associated with 3D GIS and the Web. As illustrated in the preceding examples, **geography is typically the VR environment in which users are immersed and able to navigate.** The geographic environment can be at the global, regional, or street level and requires geographic data such as elevation, ground imagery, features, and a library of 3D structural models. Such data can be cached on the server and downloaded to the client software via the Web. **The Web is more than just a means of delivering VR software and data from the server to Web clients. The Web also provides a platform that allows users to build VR collaboratively and to interact with each other in virtual communities.** For example, there are hundreds of millions of users registered to virtual worlds. The interactions and the relationships among the players are the major attraction of virtual worlds. VR has great potential as an application for many domains, including sociology, psychology, commerce, marketing, entertainment, education, and the military.

## 10.1.8 DIGITAL EARTH

Digital Earth (DE) is a visionary concept, popularized by former Vice President Al Gore. In a speech at the California Science Center in Los Angeles on January 31, 1998, Gore described DE as a multi-resolution, three-dimensional representation of the planet into which vast quantities of data can be embedded and with which citizens can interact to understand the Earth and human activities.

*"Imagine, for example, a young child going to a Digital Earth exhibit at a local museum. After donning a head-mounted display, she sees Earth as it appears from space. Using a data glove, she zooms in, using higher and higher levels of resolution, to see continents, then regions, countries, cities, and finally individual houses, trees, and other natural and man-made objects. Having found an area of the planet she is interested in exploring, she takes the equivalent of a 'magic carpet ride' through a 3D visualization of the terrain. Of course, terrain is only one of the numerous kinds of data with which she can interact. Using the system's voice recognition capabilities, she is able to request information on land cover, distribution of plant and animal species, real-time weather, roads, political boundaries, and population. She can also visualize the environmental information that she and other students all over the world have collected as part of the GLOBE project. This information can be seamlessly fused with the digital map or terrain data. She can get more information on many of the objects she sees by using her data glove to click on a hyperlink. To prepare for her family's vacation to Yellowstone National Park, for example, she plans the perfect hike to the geysers, bison, and bighorn sheep that she has just read about. In fact, she can follow the trail visually from start to finish before she ever leaves the museum in her hometown.*

*She is not limited to moving through space, but can also travel through time. After taking a virtual field trip to Paris to visit the Louvre, she moves backward in time to learn about French history, perusing digitized maps overlaid on the surface of the Digital Earth, newsreel footage, oral history, newspapers, and other primary sources. She sends some of this information to her personal e-mail address to study later. The timeline, which stretches off in the distance, can be set for days, years, centuries, or even geological epochs, for those occasions when she wants to learn more about dinosaurs."* (Gore 1998)

The vision of DE implied requirements about access to high computing power, a broadband Internet, interoperability of systems, and above all, spatial data acquisition, storage, and retrieval. These requirements seemed almost impossible to achieve in 1998 (Craglia et al. 2008). A decade later, clearly many aspects of this vision have been realized, evidenced by the popularity of virtual globes such as NASA World Wind, Google Earth, Microsoft Bing Maps 3D, and ArcGIS Explorer. Craglia and the others argued that the vision of Digital Earth has not yet been fully achieved and suggested eight key elements of the next-generation Digital Earth:

1.  Not one Digital Earth, but multiple connected globes addressing the needs of different audiences —citizens, communities, policy makers, scientists, and educators

2.  Problem oriented—for example, the environment, health, social benefits—and transparent in the impacts of technologies on the environment

3.  Allowing search through time and space to find similar/analog situations with real-time data from both sensors and people (different from what existing GIS can do, and different from adding analytical functions to a virtual globe)

4.  Asking questions about change, identifying anomalies in space in both human and environmental domains (flagging things that are not consistent with their surroundings in real time)

5.  Enabling access to data, information, services, and models as well as scenarios and forecasts from simple queries to complex analyses across the environmental and social domains

6.  Supporting the visualization of abstract concepts and data types (e.g., low income, poor health, and semantics)

7.  Based on open access and participation across multiple technological platforms and media (e.g., text, voice, and multimedia)

8.  Engaging, interactive, exploratory, and a laboratory for learning and multidisciplinary education and science

The development of Digital Earth will have a tremendous influence on mankind. It drives advancement in Earth observations and the creation of spatial data infrastructure at the global, national, and local levels. It can provide rich information to aid government decision making as well as scientific research. It plays a strategic role in addressing such societal challenges as natural disasters, natural resource depletion, energy shortages, and environmental degradation (Guo 1999; Foresman, Guo, and Fukui 2004).

A digital city is a concept derived from Digital Earth. **In its broadest sense, "digital city" is a metaphor for modernization and automation in which computers, the Internet, and GIS technology provide innovative services that meet the needs of government, government staff, citizens, and businesses.** Creating digital cities is an important initiative in many countries such as China. Hundreds of digital city projects have been implemented in China by national, provincial, and municipal governments as well as by universities. The actual implementation and functions of a digital city vary. For example, a digital city can consist of

● 2D maps with basic information listings such as an online map-based city guide. These "cities" are often created to promote tourism and for advertising purposes.

● 3D maps of the city, sometimes including buildings with different levels of realism. Users can get a better visual understanding of the city by using these maps.

● Business process systems that implement actual government workflows, facilitate collaboration among multiple agencies, and provide infrastructure and functionality support for e-government and e-business.

## 10.1.9 SENSOR WEB AND SENSOR NETWORK

There are different definitions of sensor Web. Kevin Delin of NASA is credited with coining the term in 1997 (Botts and Robin 2007). As defined in the NASA New Technology Report on Sensor Webs (Delin, Jackson, and Some 1999), **a sensor Web is a system of wireless, intracommunicating, spatially distributed sensors that are deployed to monitor and explore environments.** A sensor Web (figure 10.8) is capable of automated reasoning, because it can perform intelligent autonomous operations in uncertain environments, respond to changing environmental conditions, and carry out automated diagnosis and recovery. In this scenario, the sensors are able to communicate with each other and information gathered by one sensor is shared and used by other sensors. The sensor Web concept is closely related to the vision of the "Internet of things," in which real-world objects such as lights, cars, and packages are interlinked and connected to the Internet. In the vision of the Internet of things, the location and status of every object in the world can be tracked and the network is intelligent enough to self-organize information and automatically respond to the context, circumstances, or environment (Gershenfeld, Krikorian, and Cohen 2004). **In more general terms, a sensor Web is a Web-accessible network of sensors and archived sensor data that can be discovered and accessed using standard protocols and APIs** (Botts et al. 2006). This general definition is what Delin and Shannon Jackson (2001) called "distributed sensors" or "sensor networks," which should be distinguished from the sensor Web they defined previously. For the general discussion in this book, the two definitions are not distinguished from each other.

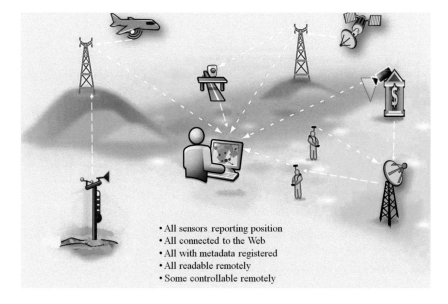

- All sensors reporting position
- All connected to the Web
- All with metadata registered
- All readable remotely
- Some controllable remotely

**Figure 10.8** The broad term sensor Web includes sensor networks that can be discovered and accessed over the Web.

Research on the sensor Web, including the OGC SWE (sensor Web enablement) suite of specifications, aims to allow users to do the following:

- Quickly discover sensors and sensor observations that can meet their needs, based on parameters such as location, observations, and quality of measurement

- Readily access sensor information in standard encoding that will allow their software to automatically process and geolocate observations

- Readily access real-time or archived time series observations in standard encoding

- Program sensors and models, when possible, to meet specific needs

- Publish, subscribe to, and receive alerts that can be issued by sensors or sensor services based on certain criteria

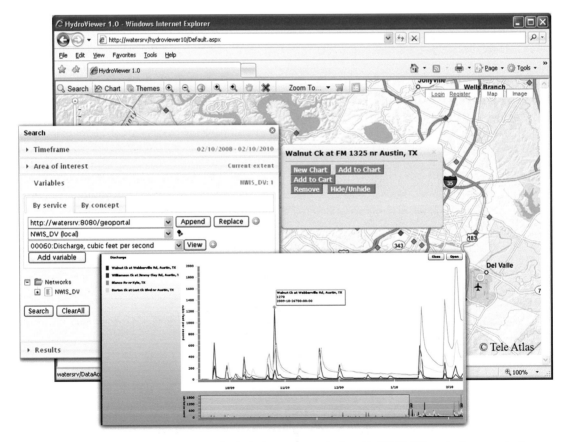

**Figure 10.9** Researchers can search online for hydrologic sensors deployed by USGS, EPA, and other organizations, retrieve historic and near real-time sensor data, chart the data, and perform hydrologic analysis.

Courtesy of Tele Atlas North America, Inc., and U.S. Geological Survey.

The sensor Web can provide timely, comprehensive, and continuous observations of the environment. As sensors get cheaper, smarter, smaller, and faster, countless sensors such as flood gauges, air pollution monitors, stress gauges on bridges, mobile heart monitors, Web cams, satellite-borne Earth imaging devices, and sensor systems are being deployed. They can sense hundreds of physical, chemical, and biological properties. The increased coverage of mobile networks allows the sensed data to be conveyed to the Web in real time. The sensor Web allows users to mash up the sensed data to build expert systems such as environmental monitoring, utility SCADA (supervisory control and data acquisition) operations, real-time detection and early warning systems, and rapid coordination of emergency response.

For example, the National Water Information System by USGS has millions of sites with sensors measuring stream flow, groundwater levels, and water quality. The STORET (storage and retrieval) system of the U.S. Environmental Protection Agency has a large number of sensors measuring water quality and other biological and physical data. The Automated Surface Observing Systems (ASOS) program is a joint effort of the National Weather Service (NWS), Federal Aviation Administration (FAA), and Department of Defense (DOD). ASOS serves as the nation's primary surface weather observing network that can support weather forecast activities and aviation operations and, at the same time, support the needs of the meteorological, hydrological, and climatological research communities. A common problem with today's sensor networks is that their data is provided in different interfaces and formats. Researchers such as CUAHSI–HIS (Consortium of Universities for the Advancement of Hydrologic Science Inc.–Hydrologic Information System) are addressing such problems by providing Web services, tools, standards and procedures that enhance access to more and better hydrologic observation data (Maidment et al. 2009). With CUAHSI HydroViewer (figure 10.9), a joint research project of CUAHSI and ESRI, scientists and the public can search for sensors by their location, the variables they measure, and the time span the measurements cover. Researchers and the public can dynamically retrieve the historical and near-real-time data, visualize it, and perform flood and other analysis with it (Ye and Djokic 2009).

## 10.1.10 CLOUD COMPUTING AND CLOUD-BASED GIS

Cloud computing, the notion that many of the computing tasks handled by individual computers locally will instead be tackled by huge computer centers connected over the Internet and available as Web-based services, is rapidly emerging as an important research area and technology trend. Taking advantage of cloud computing technology, cloud-based GIS becomes a promising business model, where vendors provide GIS software and contents to users.

### THE CONCEPT

Cloud computing has several definitions. According to the National Institute of Standards and Technology, **cloud computing is a computing paradigm for enabling on-demand network access to a shared pool of configurable computing resources (e.g., networks, servers, storage, applications, and services) that can be rapidly provided and released with minimal management effort or service provider interaction** (Mell and Grance 2009). Cloud computing (figure 10.10) can be defined by the following three characteristics (Berry 2009; Kouyoumjian 2010):

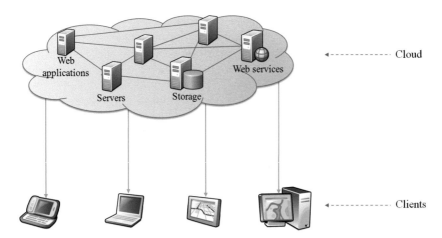

**Figure 10.10**  Cloud computing provides on-demand network access to a shared pool of configurable computing resources over the Internet.

1. **It acts as a service:** Cloud computing furnishes technological capabilities—commonly maintained off-premise—that are delivered on demand as a service via the Internet. Since a third party owns and manages public cloud services, consumers of these services do not own assets in the cloud model but pay for them on a per-use basis.

2. **It is dynamically scalable:** Instead of a static system architecture, cloud computing supports the ability to dynamically scale up and quickly scale down, offering cloud consumers high reliability, quick response times, and the flexibility to handle traffic fluctuations and demand.

3. **It involves virtualized resources:** Workloads are allocated among a large number of interconnected computers acting as a single device.

Cloud computing has three service models, and a provider may provide one or more of these services:

1. **IaaS (Infrastructure as a Service)** refers to computing power, storage, operating systems, and supporting infrastructure, for example, firewalls and load balancers. These services are delivered as off-premise, on-demand services rather than as dedicated, on-site resources. Examples include Amazon Simple Storage Service (S3) and Amazon Elastic Computing Cloud (EC2).

2. **PaaS (Platform as a Service)** provides an application platform or middleware as a service on which developers can build and deploy custom applications. Common solutions provided in this tier range from APIs and tools to database and business process management systems to security integration, all of which allow developers to build applications and run them on the infrastructure that the cloud vendor owns and maintains. Microsoft Windows Azure platform services are often referenced as PaaS solutions at this middleware tier.

3. **SaaS (Software as a Service)** comprises end-user applications delivered as a service, rather than as traditional on-premises software. SaaS applications are sometimes called Web-based software, on-demand software, or hosted software. Examples include the customer relationship management (CRM) functionalities available at `salesforce.com`.

A closely related concept to cloud computing is grid computing, a form of distributed computing where a virtual super computer is composed of a cluster of networked, loosely coupled computers acting in concert to perform very large tasks. Grid computing infrastructure such as the National Science Foundation TeraGrid provides valuable resources for many scientific problems. While both grid computing and cloud computing build virtual computer environments based on networked computers and provide virtual resources to remote users, they differ in several aspects (Pierce et al. 2009). From the user's perspective, grid computing is not easily accessible by general users since it often involves complex middleware and sophisticated batch job scheduling algorithms of parallel computing. On the other hand, cloud computing is readily available and easily accessible to Web users for many different application fields.

## THE PROS AND CONS OF CLOUD COMPUTING

Cloud computing brings a number of advantages to user organization and provides opportunities to become more cost-effective, productive, and flexible in delivering new capabilities (Kouyoumjian 2010):

- **Reduced cost:** Many services "in the cloud" are free or can be rented using the pay-as-you-go pricing model, which allows users to pay for just what they use. For most user organizations, renting cloud-based services is much cheaper than acquiring, installing, and maintaining infrastructures locally by themselves.

- **Reduced complexity:** Renting assets shifts the duty of installing and maintaining on-premises data centers to the cloud vendor, alleviating the customer's responsibility for software and hardware maintenance, ongoing operation, and support.

- **Increased availability:** Cloud computing is typically dynamically scalable, allowing it to handle large fluctuations in usage while providing reliable services to customers.

- **Increased mobility:** Inheriting the advantages of the Web, services provided by cloud computing are ubiquitously accessible from anywhere there is an Internet connection. This is a benefit for both office employees and mobile workforces.

Despite the many benefits of cloud computing, it is important to be aware of the concerns and risks of doing business in a cloud :

- **Limited contents and functions:** Cloud computing vendors provide services based on market demands and business returns. It is common for customers to find that the data and functions they need are not available from the cloud. The rapidly growing cloud architecture will provide more diverse contents and functions in the future but it won't be able to completely eliminate the problem. There are situations where customers will still have to resort to the Software plus Services (S+S) architecture (see section 7.4).

- **Security and privacy concerns:** There have been several instances of security breaches with public cloud architectures, and they should serve as reminders to be vigilant and cautious in the on-demand marketplace. One solution is using a private cloud. **Unlike a public cloud, which offers its services to the Internet, a private cloud is usually deployed on an organization's private network and operates solely for the organization's use.** A private cloud capitalizes on data security, corporate governance, and reliability, fitting the needs of many mission-critical or highly secured business interests. But a private cloud is on-site rather than being off-site and needs to be maintained by the user organization itself, thus losing such cloud computing advantages as reduced cost and complexity.

- **Lack of standards to ensure interoperability:** There are no comprehensive standards as yet to ensure free movement between cloud providers.

## CLOUD-BASED GIS

**Cloud-based GIS, or the GeoCloud, adopts cloud computing technology to provide GIS capabilities.** The notion is simple: instead of acquiring and managing GIS data and software locally, let someone else install, host, and manage it for you on remote servers. You can think of the GeoCloud as a big computer center that sits somewhere over the Internet and provides GIS capabilities to Web users. Using cloud-based GIS, you can directly use a provider's powerful infrastructure and GIS capabilities. For example, ESRI provides cloud-based GIS in multiple forms:

- **Ready-to-use content and services:** ArcGIS Online has a variety of basemap and geoprocessing services (see chapter 7). ESRI Business Analyst Online (BAO) combines GIS technology with extensive demographic, consumer, and business data to deliver on-demand business analysis, boardroom-ready reports, and maps over the Web (see chapter 8). These contents and services are provided via Web APIs as well as Web applications. ArcGIS Online and BAO products are offered in the PaaS and SaaS models.

- **Community maps program:** The ArcGIS Online content-sharing program enables users and organizations to contribute geographic data content for ESRI to host (see chapter 7). This is the IaaS service model of cloud computing.

- **ArcGIS Server running in the cloud:** ESRI provides ArcGIS Server for Amazon Machine Images (AMI) for use in Amazon cloud. This enables organizations to leverage ArcGIS in their own instances in the cloud to supplement local in-house resources and/or reduce their capital expenses for hardware, making it possible for organizations to quickly adjust the capacity of ArcGIS Server services and applications to user demand. This is a PaaS service model that offers organizations greater flexibility and capabilities than ready-to-use contents and the community maps program.

- **ArcGIS.com:** The ArcGIS Web site is a GIS platform in the cloud. It enables users to use diverse Web resources, including those offered by the ESRI ArcGIS Online cloud and those contributed by the user community, to perform a series of GIS operations, and to create their own Web GIS applications, without needing to have GIS software and data in their local IT infrastructure.

Cloud-based GIS inherits the advantages of cloud computing, such as reduced cost and complexity, quicker development times, increased availability and mobility, as well as the disadvantages of cloud computing, including security, ownership, and liability concerns. While there is a lot of hype surrounding cloud computing and cloud-based GIS, the following assessment is probably more reasonable:

*"Turning the oil tanker of GIS may take a lot longer than technical considerations suggest—so don't expect GIS to 'disappear' into the clouds just yet. But the future possibility is hanging overhead."* (Berry 2009)

## 10.2 FUTURE TRENDS

The Internet and the Web have progressed rapidly, and the pace is accelerating. The Next Generation Internet will be faster and more mobile. The future Web will be even bigger, more sociable, and more intelligent. Taking advantage of these strengths, Web GIS will thrive in the future.

### 10.2.1 A FASTER AND MORE MOBILE INTERNET

Since its inception in the 1960s, the Internet has evolved from a simple means of communication among computers to an indispensable element of our daily lives. It is a fundamental part of modern society, and the hub of a vital national and global infrastructure. It connects billions of private and business users via desktop and mobile devices all over the world and supports a multitude of applications. All these applications exert new demands on the Internet. Online photo albums, Internet phones, video-sharing sites, videoconferencing, movie downloads, and online gaming generate huge quantities of traffic and produce exponential demands on bandwidth. In addition, applications such as e-government, e-commerce, e-science, e-health, and personal communications demand greater Internet security as well.

In order to improve the Internet to accommodate current and future demands, many countries have conducted research initiatives to enhance and redesign the Internet. The United States has NGI (Next Generation Internet), GENI (Global Environment for Network Innovations), and Future Internet Design (FIND) research programs; Japan has NXGN (Next Generation Network) and NWGN (New Generation Network); and Europe has the FIRE (Future Internet Research and Experimentation) framework. Many countries are participating in GTRN (Global Terabit Research Network) to develop the Next Generation Internet (ITU 2009; Eberspächer and Tran-Gia 2008). The future Internet will be

- **Faster:** For example, the Internet2 network provides a 100 Gbit/s network backbone to more than 300 educational institutions, corporations, and nonprofit and government agencies in the United States. Researchers at Bell Labs managed to transfer optical data at the incredible rate of 2.05 Tbit/s (Gardner 2008). Such progress is bringing the 100 Gbit/s transmission speed closer to reality for general Internet users.

- **Broader with more IP addresses:** IP addresses on the current Internet are being depleted. The current Internet uses the IPv4 (Internet Protocol version 4) standard, which uses a 32-bit address and only supports 232 addresses, or about 4 billion. While still in its initial deployment, IPv6 (Internet Protocol version 6) is likely to replace IPv4, and it will support 2,128 addresses. In theory, IPv6 would allow every person in today's population to have 16 million IP addresses. IPv6 is one of the steps to making the Internet of things a reality. The implication is huge—imagine your cell phone, your car, your refrigerator, and the sensors of the sensor Web each having a unique IP address that can be remotely accessed and controlled from the Internet.

- **More mobile:** The popularity of mobile devices demands the development of mobile communication technology. While the 3G and pre-4G networks are becoming available in more countries and with broader coverage, the real 4G network, targeting 100 Mbit/s, is on its way. LTE-Advanced and WiMAX offer data speeds up to 1 Gbit/s. In terms of spatial coverage, it's possible that satellite phones will become affordable in the future, providing a stable Internet connection even to hikers in remote mountain areas.

- **More secure:** Research on the Next Generation Internet is giving full consideration to security issues in the course of architectural design, so that the controllability and manageability of network security can be greatly enhanced.

- **Greater penetration:** The Internet will be used by an even greater percentage of the world's population. Millions more people will adopt broadband Internet and mobile Internet services. More people will be connected to the Web from their home, their office, and on the road.

## 10.2.2 A SMARTER AND MORE SOCIABLE WEB

In 1990, Tim Berners-Lee laid the groundwork for what has arguably been one of the most important modern inventions, the World Wide Web. Now, we are in the era of so-called "Web 2.0." And just as there was a Web 1.0 in the beginning, there will be a Web 3.0 in the future as the Web continues to evolve. And whereas Web 1.0 was mainly about read-only content and Web 2.0 is characterized by user-generated content and the read/write Web, Web 3.0 will be mainly about the Semantic Web and intelligent search.

### WEB 2.0—WE'RE IN THE MIDST OF IT

The term "Web 2.0" is often recognized as a buzzword. It suggests a new version of the World Wide Web, but it does not refer to an update of any technical specifications. Berners-Lee criticized the term as a piece of meaningless jargon (Laningham 2006). However, the term Web 2.0 has been widely accepted and used to represent the cumulative changes in the ways software developers and end users interact with the Web.

Basically, Web 2.0 means the Web is more interactive, customizable, social, and multimedia-intensive. The principles of Web 2.0 (O'Reilly 2005), introduced in chapter 1 and reflected in other chapters, include the following:

- **Use of collective intelligence:** Examples of harnessing collective intelligence in the Web GIS context include VGI (see chapters 7 and 10) and PPGIS (see chapter 10), which encourage users to contribute data and ideas, and lightweight geospatial mashups (see chapter 4), which aid users in building their own Web applications.

- **The Web as a platform:** The Web is a platform for computing and software development. The concept has been reflected in geospatial Web services (see chapter 3) and mashups (see chapter 4), cloud computing (see chapters 7 and 10), and NSDI 2.0 (see chapter 7).

- **Lightweight programming models:** Web mapping APIs for JavaScript, Flex, and Silverlight empower users to assemble various Web sources to create new and innovative mashups (see chapter 4).

- **A rich user experience:** Web applications should be fast, easy to use, and fun. This ideal is discussed in user experience design (see chapter 2), optimizing Web services (see chapter 3), and browser-side Web APIs (see chapter 4).

- **Data as the next "inside intel":** As a metaphor, inside intel here refers to the value of unique data sources that are hard to re-create. The quality of unique data is reflected in ArcGIS.com and ArcGIS Business Analyst Online (see chapters 7 and 8), which provide a range of unique and value-added Web services. VGI (see chapters 7 and 10) also possesses unique values and can be turned into systems of services.

- **Software that runs on multiple types of devices:** Many people use desktop computers to access the Internet, but they can also use devices like smartphones and video game consoles. This is reflected in mobile GIS (see chapter 5).

## WEB 3.0—THE INTELLIGENT SEMANTIC WEB

While disputing the existence of "Web 2.0," Berners-Lee has his own vision for the future of the Web (Berners-Lee, Hendler, and Lassila 2001), and he calls it the Semantic Web (see chapter 6). Others call it Web 3.0 (Strickland 2008).

The current Web is designed so that human users can comprehend it. Computer programs such as search engines are able to scan for keywords, but they do so mostly by the spelling of the words, and not by their meaning. This can lead to problems. For example, search engines can miss Web pages that contain the information users want, resulting in information being lost in the Web because it cannot be easily discovered. This hidden Web can be much bigger than the visible Web. Another problem is information overload as search engines return numerous pages of results that can contain a lot of irrelevant information. Other problems relate to information reuse and integration, which requires people to write specific programs rather than having machines automatically process the information.

The Semantic Web will be more intelligent. In the Semantic Web, the meaning (semantics) of information on the Web will be painstakingly defined, resulting in a web of data that can be understood and processed directly by machines. One way to implement this vision is to tag each term on the Semantic Web with a Uniform Resource Identifier (URI). This URI points to the term's unambiguous meaning in ontology, which defines the relationships among a group of terms, and is present in the form of OWL (Web Ontology Language) (see chapter 6). Intelligent software can accurately understand the meaning of each term and automatically process the information. Berners-Lee provided an example in an interview:

"*For example, Berners-Lee said, a Web site that announces a conference would also contain programming with a lot of related information embedded within it.*

*A user could click on a link and immediately transfer the time and date of the conference to his or her electronic calendar. The location—address, latitude, longitude, perhaps even altitude—could be sent to his or her GPS device, and the names and biographies of others invited could be sent to an instant messenger list.*

*In other words, the 'mark-up' language behind each Web page would be cross-referenced into countless other databases, once developers agreed on a common set of definitions.*" (Shannon 2006)

The Semantic Web is a promising concept, but it is also an ambitious project with many challenges. Constructing a comprehensive ontology and tagging the terms on every Web page involves huge efforts on the part of both Web standards bodies and Web users. There are already a myriad of projects geared toward this, but it will take a long time to fully realize the Semantic Web.

### 10.2.3 TOWARD A TRUE SOCIETAL GIS

It is difficult to predict exactly how, and to what extent, Web GIS will advance, but past lessons and successes of Web GIS point to the following future trends in Web technology, architecture, and applications.

#### OPEN GEOSPATIAL WEB SERVICES

Geospatial Web services are, and will continue to be, the heart of Web GIS. They are the building blocks of mashups, a main basis of cloud-based GIS, and the foundation of NSDI 2.0. Geospatial Web services unlock GIS data and functionalities and provide a flexible way to share geospatial data and functionality. GIS professionals can serve their authoritative data and knowledge as Web services—not just map services, but geoprocessing services that fit the diverse needs of society. VGI collected and validated also can be shared as Web services. This geocollaboration will create an ecosystem of geoservices, in which new, value-added, and innovative applications will thrive. This will serve to maximize the return on geospatial investment for society as a whole. Lessons learned in the past and new technologies in the future can improve the quality of Web services in support of easy, fast, reliable, scalable, and secure applications.

#### CLIENTS THAT ARE MORE POWERFUL

Web client technology continues to advance. The competition among Web browser vendors has resulted in faster browsers in terms of HTML display, JavaScript execution, and 3D graphics rendering. HTML5 is a groundbreaking upgrade to the Web presentation specification with enticing features, including offline local data storage capability for fast performance, offline operation capabilities, enhanced multimedia and vector graphics for a rich user experience, and extended APIs. At the same time, Flex/Air and Silverlight/WPF are being enhanced to provide richer and more powerful Web clients.

The implication is that Web GIS clients will be more powerful in the future. Today's thin clients (i.e., browser-based clients) are much more powerful than the thick clients of ten years ago—a trend that will continue. The future Web client will deliver a much more expressive, intuitive, fun, and faster responding user experience than today's Web clients. Client-side, including browser-side, Web GIS APIs will have increased geoprocessing capabilities and will support client-side mashups with sophisticated functionality.

#### MOBILE AS THE PERVASIVE CLIENT PLATFORM

Mobile phones and the mobile Internet are quickly ramping up and being adopted faster than desktop computers and the desktop-based Internet. Mobile devices will likely surpass desktops and notebooks in the not too distant future as the primary platform for accessing online information.

The implication is that mobile devices will be the pervasive Web GIS client platform. GIS will be delivered to the hands of corporate field crews as well as general commercial mobile GIS users. People use location-based services for work, for pleasure, for travel, and for social networking. Augmented reality applications can supply users with information about their surroundings that they can't sense directly (see chapter 5). The information can range from what a building looked like historically to what an area will look like in a future design, or even to who the person is sitting across from you at a party. Mobile GIS can record, report, and analyze rapidly changing phenomena in situ. This real-time capability embraces the time dimension (T) and helps extend GIS from 3D (X, Y, and Z) to 4D (X, Y, Z, and T). The vision of using GIS for anything, anytime, anywhere, and by anyone will be realized, and perhaps faster than we can imagine.

## APPLICATIONS THAT GO BEYOND VISUALIZATION

The long tail theory illustrates how the economic success of vendors like Amazon.com, Netflix, and iTunes lies in selling a large variety of items, especially unique items, even if in relatively small quantities, for a significantly greater total. Applying this business model to the Web GIS market (see chapter 1), it is clear that there are significant untapped opportunities in the "long tail" of the Web GIS market. Many diverse organizations have specific needs for incorporating Web GIS into their business operations. These specialized needs demand domain-specific solutions, which are usually formulated by the organizations themselves with the aid of professional GIS software vendors and GIS consulting firms.

Domain-specific applications can involve huge volumes of data, unique business functions, and complex processes. There is plenty of room for creativity and innovation in these niche markets. The solutions will go beyond simple maps and visualization to customized geoprocessing services that deliver a wide range of functionality. The solutions will need to hide their complexity, be fast, use domain-specific language, and reflect domain-specific workflows.

## GIS DELIVERED FROM THE CLOUD

Cloud computing and cloud-based GIS have attracted much attention from both online GIS vendors and consumers. With vendors providing richer services online, more desktop users will be tempted to switch to the cloud platform. Because of the free contents available from the cloud and the affordable pay-as-you-go pricing model, cloud-based GIS will penetrate into organizations that have not used GIS before or that wouldn't be able to afford GIS on their own. While it's not realistic to expect that all needed data, maps, and functions be available from the cloud, and while there are still concerns about security, liability, and standards, cloud-based GIS will be a primary way in which GIS is delivered in the future.

## A MORE INTELLIGENT WEB GIS

The full realization of the Semantic Web is still a long way off. But according to the prediction of Berners-Lee, realization of the Semantic Web would help eliminate or reduce the limitations posed by the vagueness, uncertainty, and inconsistency of natural language. The Semantic Web could potentially bring major breakthroughs in semantic interoperability for geospatial information discovery, retrieval, data mining, and integration. Discovering appropriate Web resources can be an accurate operation. If two Web resources have different units and different coordinate systems, the unit conversion or coordinate transformation can happen automatically without the need of human intervention. Future Web GIS can be more intelligent in analyzing the data that's available over the Web, extrapolating new ideas, and fulfilling many kinds of requests.

## GIS ACTING IN REAL TIME

Many elements in Web GIS are of a real-time nature, such as the concurrent measurements from sensor Webs and sensor networks; the incidents immediately reported by field crews or citizens using mobile devices; the current locations of GPS units deployed on helicopters, fire trucks, police cars, delivery trucks, and mobile phones; the instantaneous data feeds and analysis results from GIS servers; and the synchronous information exchanged among multiple parties in scattered locations. This real-time aspect allows GIS data to be accessed, processed, analyzed, and presented in real time—a feature that is needed to solve many of today's challenges.

## CONVERGENCE OF THE VIRTUAL GLOBE AND VIRTUAL WORLD

Online virtual worlds such as Second Life are a hot topic (see section 10.1.7). There are hundreds of millions of players inhabiting these worlds as avatars. The avatars build their 3D virtual homes, interact with other avatars, and sell and purchase virtual merchandise. But these virtual worlds are mostly fantasy worlds. For these worlds to gain more practical use, they can be built on top of virtual globes—that is, the virtual version of the real world—to allow people to get the real work done in a virtual world environment (Driver and Jackson 2008).

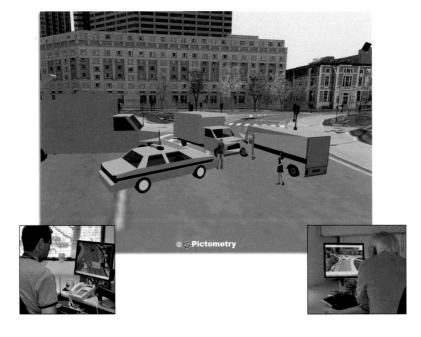

**Figure 10.11** With a virtual world implemented on top of a virtual globe, emergency response staff can take part in many types of disaster training in a virtual setting that closely resembles their real urban landscape.

Courtesy of Pictometry International Corp.

Online virtual globes already have detailed ground imagery and street maps, and they already allow users to add 3D buildings and other structures. If avatars can be added, if convenient ways to interact with them are implemented, and if rules are established for these virtual worlds, the opportunities for virtual globes can be boundless. Imagine emergency response staff taking part in disaster training drills, from auto accidents to massive explosions, in a virtual setting (figure 10.11) that is the same as their real urban landscape. Imagine that governments can develop 3D representations of plans or designs, and then meet with residents (avatars) to obtain their input. Imagine that you will be able to surf the world, navigate through a city, enter a virtual mall, examine the merchandise, and make a purchase, or find the house of a friend, knock at the door, shake hands, talk, and interact with friends using tactile feedback controls. The potential for all kinds of GIS applications is immense.

Web GIS was conceived in 1993. Looking back at the crude prototype (see figure 1.6 in chapter 1), it would be difficult, if not impossible, to imagine the enormous advances and the vast successes that comprise today's Web GIS. Similarly, it is difficult to predict the exact direction that Web GIS will take. But it is certain that the Web will be the pervasive platform for GIS in the future. Wire and wireless communications can keep you connected no matter where you are—whether you are in the office, at home, or on the go. Web GIS will be ubiquitous—available to your desktop, laptop, cell phone, GPS navigator, and other mobile devices anywhere, anytime. It will be seamlessly integrated into the operation and decision making of government, business, and all organizations, and it will become an indispensible part of most citizens' lives and jobs. While there are issues remaining and challenges ahead, Web GIS can eventually deliver a truly societal GIS that will benefit all of society.

## Study questions

1. List some hot topics related to Web GIS.
2. What is VGI? Describe some scenarios to collect VGI.
3. What is geocollaboration? Describe a situation in which remote and synchronous collaboration is needed, and provide suggestions on how to achieve such collaboration.
4. What is geotagging, and what value does it have for GIS?
5. What is geoparsing, and why is the process challenging?
6. What is geotargeting? List the methods to determine a user's location and describe the applications of geotargeting.
7. What is online VR? Describe several VR examples in the context of Web GIS.
8. What are the broad and narrow definitions of the sensor Web?
9. What are cloud computing and cloud-based GIS? What advantages and disadvantages do they have? Describe the different service models that cloud-based GIS can offer.

10. What are the main characteristics of the Next Generation Internet? How will it influence the future of Web GIS?

11. What is the Semantic Web? What implications does it have for Web GIS?

12. What are the main trends of future Web GIS development?

13. Imagine that there is a virtual world integrated with a virtual globe. Describe some scenarios in which you, your friends, your local businesses, and your local government can use the virtual world.

## References

Batty, Michael. 2007. Planning support systems: Progress, predictions, and speculations on the shape of things to come. CASA Working Papers 122. Centre for Advanced Spatial Analysis, University College London. http://eprints.ucl .ac.uk/15175/1/15175.pdf (accessed December 28, 2009).

Berners-Lee, Tim, James Hendler, and Ora Lassila. 2001. The Semantic Web. *Scientific American*, May. http://www.sciam .com/article.cfm?articleID=00048144-10D2-1C70-84A9809EC588EF21 (accessed November 11, 2009).

Berry, Joseph. 2009. GIS and the cloud-computing conundrum. *GeoWorld*, September. http://www.geoplace.com/ME2/ dirmod.asp?sid=&nm=&type=MultiPublishing&mod=PublishingTitles&mid=13B2F0D0AFA04476A2ACC02ED28A405F&tier=4 &id=7F932400DCF44F8E9ACF43A7536A77D5 (accessed December 25, 2009).

Botts, Mike, and Alex Robin. 2007. Bringing the sensor Web together. *Geosciences*, October. http://www.brgm.fr/ dcenewsFile?ID=473 (accessed December 20, 2009).

Botts, Mike, George Percivall, Carl Reed, and John Davidson. 2006. *OGC sensor Web enablement: Overview and high-level architecture*, ed. Carl Reed. http://portal.opengeospatial.org/file s/?artifact_id=25562 (accessed December 16, 2009).

Brewer, I., A. M. MacEachren, H. Abdo, J. Gundrum, and G. Otto. 2000. Collaborative geographic visualization: Enabling shared understanding of environmental processes. IEEE Information Visualization Symposium, Salt Lake City, Utah, Oct. 9–10, 137–41.

Caldwell, Douglas. 2009. Geoparsing maps the future of text documents. *Directions Magazine*, September. http://www.directionsmag.com/article.php?article_id=3268 (accessed December 23, 2009).

Carmagnola, Francesca, and Federica Cena. 2009. User identification for cross-system personalization. *Information Sciences* 179 (1–2): 16–32.

Carmichael, Richie. 2009. Geochat for Silverlight. *The Sandpit*, blog (May 15). http://kiwigis.blogspot.com/2009/05/ geochat-for-silverlight.html (accessed December 16, 2009).

Craglia, Max, Michael F. Goodchild, Alessandro Annoni, Gilberto Camara, Michael Gould, Werner Kuhn, David Mark, Ian Masser, David Maguire, Steve Liang, and Ed Parsons. 2008. Next-generation Digital Earth: A position paper from the Vespucci Initiative for the Advancement of Geographic Information Science. *International Journal of Spatial Data Infrastructures Research* 3: 146–67.

Delin, Kevin A., and Shannon P. Jackson. 2001. The sensor Web: A new instrument concept. The International Society for Optical Engineering Symposium on Integrated Optics, San Jose, Calif., January 20–26. http://www.sensorwaresystems.com/historical/resources/sensorweb-concept.pdf (accessed December 22, 2009).

Delin, Kevin A., Shannon P. Jackson, and Raphael R. Some. 1999. Sensor Webs. NASA Tech Briefs, October 1. http://www.nasatech.com/Briefs//Oct99/NPO20616.html (accessed December 22, 2009).

Driver, Erica, and Paul Jackson. 2008. Getting real work done in virtual worlds. Cambridge, Mass.: Forrester Research. `https://docs.google.com/viewer?a=v&pid=sites&srcid=ZGVmYXVsdGRvbWFpbnxtdW5pcZ292MjB8Z3g6MzJhMWNhYTkyMzQxZT g3MQ` (accessed December 28, 2009).

Eberspächer, Jörg, and Phuoc Tran-Gia. 2008. Next generation Internet. *Information Technology* 50 (6). `http://www.atypon-link.com/OLD/doi/pdf/10.1524/itit.2008.9053` (accessed December 20, 2009).

Foresman, Timothy W., Huadong Guo, and Hiromichi Fukui. 2004. Progress with the Digital Earth global infrastructure. 7th Global Spatial Data Infrastructure Conference, Bangalore, India, February 2–6.

Gardner, W. David. 2008. Researchers transmit optical data at 16.4 Tbps. *InformationWeek*, February 27. `http://www.informationweek.com/news/telecom/showArticle.jhtml?articleID=206900603&cid=RSSfeed_IWK_All` (accessed December 25, 2009).

Gellman, Robert. 2002. Privacy, consumers, and costs: How the lack of privacy costs consumers and why business studies of privacy costs are biased and incomplete. `http://epic.org/reports/dmfprivacy.pdf` (accessed January 2, 2010).

Gershenfeld, Neil, Raffi Krikorian, and Danny Cohen. 2004. The Internet of things. *Scientific American* 291 (4):76–81.

Goodchild, Michael F. 2007a. Citizens as sensors: The world of volunteered geography. *GeoJournal* 69:211–21.

———. 2007b. Citizens as voluntary sensors: Spatial data infrastructure in the world of Web 2.0. *International Journal of Spatial Data Infrastructures Research* 2:24–32.

Goodchild, Michael F., Pinde Fu, and P. M. Rich. 2007. Geographic information sharing: The case of the Geospatial One-Stop Portal. *Annals of Association of American Geographers* 97 (2): 249–65.

Gore, Al. 1998. The Digital Earth: Understanding our planet in the 21st century. `http://www.isde5.org/al_gore_speech.htm` (accessed December 18, 2009).

Guo, Huadong. 1999. Building up an Earth-observing system for Digital Earth. In *Towards Digital Earth: Proceedings of the International Symposium on Digital Earth*. Beijing, China: Science Press. `http://www.digitalearth-isde.org/cms/upload/2007-04-30/1177882766990.pdf` (accessed December 22, 2009).

ITU. 2009. The future Internet. ITU-T Technology Watch Report 10. `http://www.itu.int/dms_pub/itu-t/oth/23/01/T230100000A0001PDFE.pdf` (accessed December 25, 2009).

Kouyoumjian, Victoria. 2010. The new age of cloud computing and GIS. *ArcWatch*, January. `http://www.esri.com/news/arcwatch/0110/feature.html` (accessed January 10, 2010).

LaFramboise, Allan. 2009. ArcGIS Explorer and geo-service integration: Twitter add-in. ArcGIS Explorer Blog, December 15. `http://blogs.esri.com/Info/blogs/arcgisexplorerblog/archive/2009/12/15/arcgis-explorer-and-geo-service-integration-twitter-add-in.aspx` (accessed December 16, 2009).

Laningham, Scott. 2006. DeveloperWorks Interviews: Tim Berners-Lee. IBM podcast series, August 22. `http://www.ibm.com/developerworks/podcast/dwi/cm-int082206txt.html` (accessed April 18, 2009).

MacEachren, Alan M. 2000. Cartography and GIS: Facilitating collaboration. *Progress in Human Geography* 24 (3): 445–56.

MacEachren, Alan. M., Isaac Brewer, and Erik Steiner. 2001. Geovisualization to mediate collaborative work: Tools to support different-place knowledge construction and decision making. Proceedings of the 20th International Cartographic Conference, Beijing, China, August 6–10, 2533–39.

Maidment, David R., Richard P. Hooper, David G. Tarboton, and Ilya Zaslaksky. 2009. Accessing and sharing data using CUAHSI water data services. In *Hydroinformatics in hydrology, hydrogeology, and water resources*, ed. Ian Cluckie, Yangbo Chen, Vladan Babovic, Lenny Konikow, Arthur Mynett, Siegfried Demuth, and Dragan A. Savic, IAHS Publ. 331, 213–23. Proceedings of Symposium JS4, Hyderabad, India, September. `http://iahs.info/redbooks/a331/iahs_331_0213.pdf` (accessed January 10, 2010).

Mell, Peter, and Tim Grance. 2009. The NIST definition of cloud computing. National Institute of Standards and Technology. `http://csrc.nist.gov/groups/SNS/cloud-computing/cloud-def-v15.doc` (retrieved November 26, 2009).

Nash, Andrew. 2009. Web 2.0 applications for improving public participation in transport planning. `http://www.andynash.com/nash-publications/2009-Nash-Web2forPT-14nov09.pdf` (accessed December 9, 2009).

O'Reilly, Tim. 2005. What is Web 2.0. O'Reilly Network, September 30. `http://www.oreillynet.com/pub/a/oreilly/tim/news/2005/09/30/what-is-web-20.html` (accessed June 28, 2009).

Pierce, Marlon E., Geoffrey C. Fox, Yu Ma, and Jun Wang. 2009. Whither spatial cyberinfrastructure? Community Grids Laboratory, Indiana University. `http://grids.ucs.indiana.edu/ptliupages/publications/PNAS-Clouds-Final.pdf` (accessed December 17, 2009).

Ridley, Randy, John-Henry Gross, and John Frank. 2005. Can geography rescue text search? *ArcUser Online*, April–June. `http://www.esri.com/news/ArcUser/0405/textsearch1of2.html` (accessed December 23, 2009).

Schafer, Wendy A., Craig H. Ganoe, Lu Xiao, Gabriel Coch, and John M. Carroll. 2005. Designing the next generation of distributed, geocollaborative tools. *Cartography and Geographic Information Science* 32.

Shannon, Victoria. 2006. A "more revolutionary" Web. *New York Times*, May 23. `http://www.nytimes.com/2006/05/23/technology/23iht-web.html?_r=1` (accessed December 26, 2009).

Sieber, Renee. 2006. Public participation geographic information systems: A literature review and framework. *Annals of the Association of American Geographers* 96 (3): 491–507.

Steuer, Jonathan. 1993. Defining virtual reality: Dimensions determining telepresence. Social Responses to Communication Technologies paper 104. `http://www.cybertherapy.info/pages/telepresence.pdf` (accessed December 22, 2009).

Strickland, Jonathan. 2008. How Web 3.0 will work. HowStuffWorks.com., March 3. `http://computer.howstuffworks.com/web-30.htm` (accessed December 26, 2009).

Ye, Zichuan, and Dean Djokic. 2009. What is needed to support an integrated geo-temporal system based on CUAHSI experience. OGC Technical Committee meeting, Cambridge, Mass., June 22–26.

# LIST OF ABBREVIATIONS

**1G** — first generation of wireless phone technology

**2G** — second generation of wireless phone technology

**2.5G** — stepping stone between second and third generations

**3G** — third generation of wireless phone technology

**4G** — fourth generation of wireless phone technology

**ADL** — Alexandria Digital Library

**AEGIS** — Advanced Emergency GIS

**A-GPS** — assisted GPS

**AJAX** — Asynchronous JavaScript and XML

**ALLHAZ** — All Hazards Emergency Operations Management System

**AMF** — Action Message Format

**AMI** — Amazon Machine Images

**ANZLIC** — Australia and New Zealand Land Information Council

**AOA** — angle of arrival

**API** — application programming interface

**AR** — augmented reality

**ARPA** — Advanced Research Project Agency (part of U.S. Department of Defense)

**ARPANet** — predecessor of the global Internet

**ARRA** — American Recovery and Reinvestment Act

**ASDD** — Australian Spatial Data Directory

**ASDI** — Australian Spatial Data Infrastructure

**ASOS** — Automated Surface Observing Systems

**AVS** — Address Verification Service

**B2B** — business-to-business

**B2C** — business-to-customer

**B2E** — business-to-employee

**B2G** — business-to-government

**BA Server** — ArcGIS Business Analyst Server

**BAO** — Business Analyst Online

**BI** — business intelligence

**BLM** — Bureau of Land Management

**BRA** — Boston Redevelopment Authority

**B/S** — browser/server

**Caltrans** — California Department of Transportation

**CARE** — Citizen Active Response Effort

**CAS** — Chinese Academy of Sciences

**CDC** — Centers for Disease Control and Prevention

**CD-ROM** — compact disc read-only memory

**CERN** — European Organization for Nuclear Research

**CGDI** — Canadian Geospatial Data Infrastructure

**CGIS** — Canada Geographic Information System

**CHP** — California Highway Patrol

**COO** — cell of origin

**COP** — common operating picture

**CORBA** — common object request broker architecture

**CPU** — central processing unit

**CRA** — Community Reinvestment Act

**CRM** — customer relationship management

**CRUD** — create, read, update, and delete

**C/S** — client/server

**CSCW** — computer-supported cooperative work

**CSDGM** — Content Standard for Digital Geospatial Metadata

**CSS** — Cascading Style Sheet

**CSV** — comma-separated values

**CSW** — Catalog Service for the Web

**CUAHSI–HIS** — Consortium of Universities for the Advancement of Hydrologic Science Inc.–Hydrologic Information System

**DBMS** — database management system

**DCI** — data-collecting infrastructure

**DCOM** — Distributed Component Object Model

**DE** — Digital Earth

**DHTML** — dynamic HTML

**DOD** — Department of Defense

**DOE** — Department of Energy

**DRM** — digital rights management

**DVD** — digital versatile disc

**E-911** — enhanced 911

**EC2** — Elastic Computing Cloud

**ECMA** — European Computer Manufacturers Association

**EcoServ** — Ecosystem Service Modeling System

**EOC** — emergency operations center

**E-OTD** — enhanced observed time difference

**EPA** — Environmental Protection Agency

**ERP** — enterprise resource planning

**ETL** — extract, transform, and load

**EXIF** — exchangeable image file

**FAA** — Federal Aviation Administration

**FEMA** — Federal Emergency Management Agency

**FGDC** — Federal Geographic Data Committee

**FIND** — Future Internet Design

**FIRE** – Future Internet Research and Experimentation

**FTP** — File Transfer Protocol

**GBF/DIME** — Geographic Base File/Dual Independent Map Encoding

**GCMD** — Global Change Master Directory

**GEMET** — General Multilingual Environmental Thesaurus

**GENI** — Global Environment for Network Innovations

**GEO** — Group on Earth Observations

**GeoDesign** — methodology for geographic planning and decision making

**GEOSS** — Global Earth Observation System of Systems

**GeoWeb** — geospatial Web

**GHG** — greenhouse gas

**GIF** — Graphics Interchange Format

**GIS** — geographic information system(s) or science

**GLONASS** — Global Navigation Satellite System (Russia)

**GML** — Geography Markup Language

**GNIS** — Geographic Names Information System

**GNSS** — Global Navigation Satellite System

**GOS** — Geospatial One-Stop (portal)

**GPS** — Global Positioning System

**GRASS** —Geographic Resources Analysis Support System

**GSA** — General Services Administration

**GSDI** — Global Spatial Data Infrastructure

**GSDIA** — Global Spatial Data Infrastructure Association

**GSRP** — Geographic Search and Referencing Products

**GTRN** — Global Terabyte Research Network

**HASC** — Hospital Association of Southern California

**HDF-EOS** — Hierarchical Data Format–Earth Observation System

**HTML** — Hypertext Markup Language

**HTTP** — Hypertext Transfer Protocol

**HTTPS** — Secure Hypertext Transfer Protocol

**HUD** — Department of Housing and Urban Development

**IaaS** — Infrastructure as a Service

**ICT** — information and communication technologies

**IE** — Internet Explorer

**IFRIS** — Integrated Forest Resource Information System (Virginia Department of Forestry)

**IGSNRR** — Institute of Geographic Sciences and Natural Resources Research

**IM** — instant messaging

**INSPIRE** — Infrastructure for Spatial Information in Europe

**IP** — Internet Protocol

**IPv4** — Internet Protocol version 4

**IPv6** — Internet Protocol version 6

**IRC** — Internet Relay Chat

**ISO** — International Organization for Standardization

**ISO/TC 211** — International Organization for Standardization Technical Committee 211

**ISP** — Internet service provider

**IT** — information technology

**ITFA** — Internet Tax Freedom Act

**ITU** — International Telecommunication Union

**Java ME** — Java Mobile Edition

**JAXGIS** — GIS division of the City of Jacksonville, Florida, Information Technology Department

**JPEG** — Joint Photographic Experts Group

**JRE** — Java runtime environment

**JSF** — JavaServer Faces

**JSON** — JavaScript Object Notation

**JSP** — JavaServer Pages

**KBps** — kilobytes per second

**KML** — Keyhole Markup Language

**KMZ** — Keyhole Markup Language zipped files

**KWMIP** — Kentucky Watershed Modeling Information Portal

**LAN** — local area network

**LBS** — location-based service

**LIMS** — laboratory information management system

**LIS** — land information system

**LLUMC** — Loma Linda University Medical Center

**LS&CO** — Levi Strauss & Co.

**LTE** — long-term evolution

**MBps** — megabytes per second

**MIME** — Multipurpose Internet Mail Extensions

**MMS** — Multimedia Messaging Service

**MXD** — ESRI map file extension

**MXML** — an XML-based user interface markup language, often used in Adobe Flex

**MSD** — map service definition

**NASA** — National Aeronautics and Space Administration

**NCGIA** — National Center for Geographic Information and Analysis

**NDFD** — National Digital Forecast Database

**NEO** — NASA Earth Observations

**NEPA** — National Environmental Policy Act

**NGDC** — National Geospatial Data Clearinghouse

**NGI** — Next Generation Internet

**NGICC** — National Geospatial Information Coordination Committee

**NILS** — National Integrated Land System

**NITF** — National Imagery Transmission Format

**NOAA** — National Oceanic and Atmospheric Administration

**NORAD** — North American Aerospace Defense Command

**NPS** — National Park Service

**NSDI** — National Spatial Data Infrastructure

**NSIDC** — National Snow and Ice Data Center

**NSII** — National Spatial Information Infrastructure

**NWGN** — New Generation Network

**NWS** — National Weather Service

**NXGN** — Next Generation Network

**OAI** — Open Archives Initiative

**OGC** — Open Geospatial Consortium

**OLAP** — online analytical processing

**OMB** — Office of Management and Budget

**OpenLS** — OpenGIS Location Services

**OWL** — Web Ontology Language

**OWS** — OGC Web Services

**P2P** — peer-to-peer

**P3P** — Platform for Privacy Preferences

**PaaS** — Platform as a Service

**PARC** — Palo Alto Research Center

**PC** — personal computer

**PCA** — principal components analysis

**PDA** — personal digital assistant

**PDC** — Pacific Disaster Center

**PII** — personally identifiable information

**PLSS** — Public Land Survey System

**PNG** — Portable Network Graphics format

**POI** — point of interest

**POX** — Plain Old XML

**PPGIS** — public participation GIS

**QA** — quality assurance

**QC** — quality control

**QoS** — quality of service

**RDF** — Resource Description Framework

**ReddiNet** — Rapid Emergency Digital Data Information Network

**REST** — Representational State Transfer

**RFID** — radio frequency identification

**RIA** — rich Internet application

**RSS** — Really Simple Syndication or Rich Site Summary

**S3** — simple storage service

**SaaS** — Software as a Service

**SAS** — sensor alert service

**SCADA** — supervisory control and data acquisition

**SDI** — spatial data infrastructure

**SDK** — software development kit

**SLD** — styled layer descriptor
**SMS** — Short Message Service
**SMTP** — Simple Mail Transfer Protocol
**SOA** — service-oriented architecture
**SOAP** — Simple Object Access Protocol
**SOC** — server object container
**SOM** — server object manager
**SOS** — sensor observation service
**SPS** — sensor planning service
**S+S** — Software plus Services
**SSA** — Social Security Administration
**SSL** — Secure Sockets layer
**STORET** — storage and retrieval
**SVG** — scalable vector graphics
**SWE** — sensor Web enablement
**TDOA** — time difference of arrival
**TIFF** — Tagged Image File Format
**TIGER** — Topologically Integrated Geographic Encoding and Referencing (U.S. Census Bureau)
**TOA** — time of arrival
**UE** — user experience (also UX)
**UGC** — user-generated content
**UMB** — Ultra Mobile Broadband
**URI** — Uniform Resource Identifier
**URL** — Uniform Resource Locator
**USACE** — U.S. Army Corps of Engineers
**USDA** — U.S. Department of Agriculture
**USFS** — USDA Forest Service
**USGS** — U.S. Geological Survey
**USPS** — U.S. Postal Service
**UX** — user experience (also UE)
**VDEM** — Virginia Department of Emergency Management
**VEOC** — Virginia Emergency Operations Center
**VGI** — volunteered geographic information
**VIPER** — Virginia Interoperability Picture for Emergency Response
**VPN** — virtual private network
**VR** — virtual reality
**VRML** — Virtual Reality Modeling Language
**W3C** — World Wide Web Consortium
**WAF** — Web-accessible folders
**WAP** — Wireless Application Protocol
**WCS** — Web Coverage Service
**WFS** — Web Feature Service

**WFS-T** — Transactional WFS

**WGS 84** — World Geodetic System 1984

**WHO** — World Health Organization

**WIB** — wireless Internet browser

**Wi-Fi** — wireless fidelity

**WiMAX** — Worldwide Interoperability for Microwave Access

**WLAN** —wireless local area network

**WML** — Wireless Markup Language

**WMS** — Web Map Service

**WMTS** — Web map tile service

**WNV** — West Nile virus

**WOA** — Web-oriented architecture

**WPF** — Windows Presentation Foundation

**WPS** — Web Processing Service

**WSDL** — Web Service Description Language

**WWW** — World Wide Web

**X3D** — extensible 3D

**XAML** — Extensible Application Markup Language

**XHTML** — Extensible HTML

**XHTML MP** — Extensible HTML Mobile Profile

**XML** — Extensible Markup Language

**XMP** — Extensible Metadata Platform

# INDEX